Energy-Harvesting Materials

David J. Fisher

Copyright © 2025 by the authors

Published by **Materials Research Forum LLC**
Millersville, PA 17551, USA

Published as part of the book series
Materials Research Foundations
Volume 177 (2025)
ISSN 2471-8890 (Print)
ISSN 2471-8904 (Online)

Print ISBN 978-1-64490-366-7
ePDF ISBN 978-1-64490-367-4

Distributed worldwide by

Materials Research Forum LLC
105 Springdale Lane
Millersville, PA 17551
USA
http://www.mrforum.com

Printed in the United States of America
10 9 8 7 6 5 4 3 2 1

Table of Contents

Introduction 1

Vibration and Impact Sources 3

   Triboelectric harvesters...3

   Piezoelectric harvesters...18

   Waste-heat harvesting 61

   Miscellaneous..78

Salinity-Harvesting..94

About the Author...103

References..104

Introduction

Although the term 'harvesting' is conventionally used in this context, 'recuperation' is perhaps more apt because the former term would encompass the use of wind, solar, tidal and similar sources when what is really meant is, as it were, the gleaning of 'second-hand' energy. That is to say, it is the harvesting of energy which is the result of some primary activity and would otherwise go to waste. An early example would be the self-winding wrist-watch, an 18th-century innovation, in which the everyday activities of the wearer store energy in the watch-spring and thus power the article. The energies involved are relatively trivial, and any reciprocal effect upon the wearer goes unnoticed.

An early example of more robust energy recuperation was that of regenerative braking. With the advent of electrically-driven street-cars for public transportation, for example, it was realised[1] that a considerable amount of energy could be conserved by braking a vehicle by using some sort of generator. The energy which would otherwise have been wasted by a friction-process, which merely generated heat that was then lost to the environment, could instead be fed back into the electrical system. This concept of 'regenerative braking' is now widespread[2].

Another aspect of vehicular motion which can be exploited is of course the action of shock-absorbers. Simple springs or dashpots provide the required cushioning but, again, the energy merely ends up as heat. It is much better to, for example, use some sort of induction device in which a magnet is driven through a coil. That movement generates a useful current in the wire, while Lenz's Law provides a cushioning effect by resisting the movement of the magnet. The same idea is often used in electric torches: a magnet which can oscillate within a coil can generate power to be stored in a battery. So such a torch can be left in a vehicle and, thanks to repeated disturbances occurring during travel, will always be ready for emergency use without having to remember to replace batteries.

It must be noted however that the concept of recuperating energy from vehicular movement has sometimes been taken too far. It was not long before inventors proposed the installation of rollers or ramps in public roads[3]. The inventors argued that they were harvesting waste energy. The problem here of course is that the energy is not really being wasted; it is being used to overcome rolling resistance and atmospheric drag and such rollers become just another burden. Any energy which is generated by such rollers or ramps is obviously a resource which has simply been stolen from the road-user. Not missing a trick, it has also been proposed[4] that weighbridge-like devices should be installed at toll-booths. As the platform sinks under the weight of a vehicle, energy is generated. But the vehicle will then have to climb a ramp in order to regain its original elevation. This again is tantamount to theft … unless the 'donated' energy replaces a conventional monetary toll.

Another scheme[5] was to install windmill-like devices next to railway-lines or airport runways in order to benefit from the wind of passing trains or aircraft. Again there would be a drag on the trains or aircraft, thus requiring the use of extra fuel; which woul benefit only the windmill-owner. This scheme was publicised by unthinking journalists[6]. Locomotives travelling through tunnels have also been targeted[7].

It often seems that some inventors and designers have never heard of Newton's third law, or of the 'no free lunch' principle. One inventor even eliminated the 'middle-man', in energy-harvesting using roads, and proposed that automobiles could simply drive around a circular track containing ramps and thereby function as a power-station. This would have turned gasoline into electricity, but via a quite ridiculous process: a process which moreover was even claimed to be extracting 'gravitational energy'[8]. This highly dubious concept was however explored in some detail[9].

It is nevertheless a factor which must be kept in mind: that any energy-harvesting device should recuperate that energy without exploiting or inconveniencing the user. For instance, putting devices in sidewalks so as to power the streetlights is a good idea, provided that pedestrians do not feel as though they are walking through marshland. Installing devices in the clothing of military personnel might provide useful energy for body-worn electronic hardware, but it would be counterproductive if it fatigued the soldier to an unacceptable degree. It has been suggested[10] that breathing, chest expansion, blood-pressure changes, limb and finger motions plus footfalls could provide energy. That would seem to be a trifle tiring. But the ambitions of inventors know no bounds, leading to suggestions that energy might be harvested from prosthetic limbs[11] or even from eye movements, via contact lenses[12].

There has been a puzzling enthusiasm for the use of pendula in energy-harvesting[13]. It is clear that a pendulum which is mounted in a vehicle can exploit the accelerations and decelerations of that vehicle. It is less clear whether that energy would compensate the user for the extra fuel cost resulting from transporting the pendulum. It is puzzling moreover that such pendula[14] are often depicted as standing in fields, and it is not clear – decidedly vague indeed - what exactly is maintaining the pendulum in motion.

That is why the examples of energy-harvesting which are described in the present work are ones which have only imperceptible effects upon the user, or which involve large-scale phenomena where individuals are not involved. The former choice has come to centre largely on the use of nano-materials, piezoelectricity and triboelectricity to harvest the energy involved in a wide variety of everyday activities that involve vibrations, impacts, etc. But even in the field of piezoelectric energy-harvesting, it was once found to be necessary to point out[15] that mere rotation of a device cannot generate energy.

Vibration and Impact Sources

Triboelectric harvesters

A 3-dimensional triboelectric-electromagnetic hybrid nanogenerator was based[16] upon a magnetic ball and polystyrene spherical shells. External excitations could be transformed into relative motions between various components. The latter, comprising polystyrene spherical shells and a magnetic ball, transformed 3-dimensional mechanical energy into electrical energy via friction-electrification and electromagnetic induction. Experiments showed that the performance of devices could be affected by the direction of external vibration and by the excitation frequency. The output increased as the excitation-frequency increased. The maximum output power of the triboelectric nanogenerator was 18μW with an external load resistance of 200MΩ. The maximum output power of the electromagnetic generator was 640μW with an external load resistance of 1000Ω.

A ferrofluid-based vibration-energy harvester with a hybrid triboelectric plus electromagnetic operating mechanism was proposed[17]. A ferrofluid suspension was made by using water and magnetic nanoparticles. Electrostatic induction between a polymer side-wall and the water activated a triboelectric generator component while electromagnetic induction between the suspended nanoparticles and an outer coil activated an electromagnetic generator component. The ferrofluid-embedded hybrid energy-harvester then exhibited an extremely low threshold amplitude and a wide operating-frequency range. This feature was particularly useful for harvesting faint and irregular vibrations.

A tough adhesive anti-freezing ionic double-network conductive hydrogel was prepared[18] by choosing glucan as a first network and Fe^{3+}-COO- metal coordination cross-linked poly(2-Hydroxyethyl acrylate-co-acrylic acid) as a second network. These were cross-linked by hydrogen bonds in the presence of hydrated lithium ions, $[Li(H_2O)_n]^+$. The hydrogel offered a toughness of $16.26MJ/m^3$, a tensile stress of 4.15MPa, a tensile strain of 1213% and a compressive stress of 2.98MPa under 80% compression. The conductivity was 0.43S/m at -18C. Triboelecric nanogenerators which were based on the hydrogel furnished an electrical output of 182.42V at 25C and a good fatigue resistance. The output was 163.58V following a month of cryostorage. The device could be used for energy-harvesting at -18C.

A flexible triboelectric nanogenerator was prepared[19] from a highly porous cellulose nanofibril and polyethylenimine aerogel film which was combined with polyvinylidene fluoride nanofibre mats having excellent triboelectric outputs. Modification with polyethylenimine improved the mechanical properties of the aerogel but increased the power density by 14.4 times due to an enhanced tribopositivity. The triboelectric behaviour was further improved by using multiple layers of polyvinylidene fluoride mats. The triboelectric nanogenerator exhibited an 18.3 times improvement in output voltage and a 97.6 times improvement in power density over those found for one layer of polyvinylidene fluoride mat. The peak output power-density attained 13.3W/m^2 for a load resistance of 106Ω.

Traditional triboelectric nanogenerators with solid frictional layers have limitations liquid-solid versions have evolved which can capture ultra-low vibrational signals due to their fluidity[20]. The liquid-solid generators nevertheless offer only lower energy-conversion efficiencies when compared with solid counterparts, and also cannot be used in hybrid energy-harvesting devices. The use of ferrofluids solves some of these problems when the liquid component is replaced by a ferrofluid. Such systems can be used for hybrid energy-harvesting.

The ferroelectric and piezoelectric energy-harvesting behaviour of a 3d metal ion-containing A_4BX_6 organic-inorganic hybrid salt, $[Ph_3MeP]_4[Ni(NCS)_6]$ was studied[21]. Hysteresis-loop studies revealed a remnant ferroelectric polarization of $18.71\mu\text{C/cm}^2$, at room temperature. Composite thermoplastic polyurethane devices which contained 5, 10, 15 or 20wt% of $[Ph_3MeP]_4[Ni(NCS)_6]$ were used for piezoelectric energy-harvesting. A maximum output voltage of 19.29V and a power-density of 2.51mW/cm^3 were obtained from the 15wt% $[Ph_3MeP]_4[Ni(NCS)_6]$ device. Capacitor-charging tests of this device revealed stored energies and charges of $198.8\mu\text{J}$ and $600\mu\text{C}$, respectively.

A bistable piezoelectric-triboelectric hybrid generator possessing low potential energy barrier characteristics and low-frequency broadband energy-harvesting abilities was proposed[22]. The device consisted of 2 flexible beams and 2 rigid links which were connected by a spring. There was a positive correlation between displacement and output voltage. Adjusting the length of the rigid links could effectively increase the deformation of the piezoelectric beam and the relative displacement of the triboelectric layers, thus improving the output. A triboelectric nano-generator exhibited an energy-harvesting ability, for frequencies 3.5 to 6.5Hz, which offered a voltage of 1050V and a power of 1.84mW when subjected to an excitation amplitude of 17.5mm and a frequency of 5.5Hz. A piezoelectric generator exhibited an output at frequencies of 4 to 7Hz, which offered a voltage of 6.2V and a power of $12.8\mu\text{W}$ for an excitation frequency of 7Hz and an amplitude of 17.5mm. The bistable piezoelectric-triboelectric hybrid generator offered an improved output over a frequency range of 3.5 to 6.5Hz.

The preferentially-oriented low-dimensional layered all-inorganic halide perovskite material, $CsPb_2Br_5$, exhibits piezoelectric and ferroelectric behaviour at room temperature. A composite was created[23] by combining $CsPb_2Br_5$ microplates with polyvinylidene fluoride and used in a nanogenerator for energy-harvesting. The pristine tetragonal $CsPb_2Br_5$ microplates had a piezoelectric coefficient of 72pm/V and a remanent polarization of $0.06\mu C/cm^2$. Four composite devices, containing various weight percentages of perovskite, were compared with a pure polyvinylidene fluoride nanogenerator. The incorporation of $CsPb_2Br_5$ into the polyvinylidene fluoride matrix permitted high crystallinity and electroactive-phase nucleation of some 92% in the polyvinylidene fluoride; greater than that for pristine polyvinylidene fluoride. Under periodic vertical compression, the best of the devices offered an instantaneous output voltage of 200V, a current of 2.8µA and a power of 45µW across a 5MΩ resistor.

A review of energy-harvesting using ferroelectric materials focussed[24] on porous ferroelectric ceramics with piezo- and pyro-electric properties. The benefits of porosity in ferroelectrics such as lead zirconate titanate were particularly relevant to energy harvesting from ambient vibrations and temperature fluctuations. Vibration energy-harvesting was reviewed[25] with the aim of finding scaling-laws for known harvesting devices which were based upon electromagnetic transduction. The power density was related to length, mass, frequency and drive acceleration. The scaling-laws which were developed suggested that there was an upper limit to the power-density which was possible by using current harvesting techniques of the time. The aim was to identify the approximate scaling behaviour and upper bounds on power density and other parameters. The upper boundaries were defined by making linear fits to the edge of distributions of points on a log-log graph and expressing the result in the form,

$$parameter = 10^a x^b$$

where 10^a was the constant of proportionality and x was a scaling length or proof mass and b was a power-law exponent (table 1). The scaling laws, especially power-density, when normalized with respect to the square of the resonant frequency suggested that there is an upper limit to the power density. This upper limit led to the estimated maximum power being given by:

$$P_{max} = 1.9 \times 10^{-6} V^2 f^2$$

for a device with a volume of $V(cm^3)$ and a resonant frequency of f(Hz).

Table 1. Scaling laws for the upper power limit of electromagnetic energy harvesters

Correlation	a	b	R	SD
m versus L	2.77657	-2.20446	0.99981	0.03676
P versus L	4.27655	-2.19145	0.99558	0.20407
P versus m	0.86096	1.56501	0.99971	0.04698
PD versus L	1.2139	-2.14146	0.9704	0.18483
PD versus m	1.06663	0.92135	0.99885	0.09645
PD/g^2 versus L	0.0568	-0.78103	0.77059	0.05247
PD/g^2 versus m	0.04864	-0.69372	0.86592	0.06636
PD/f^2 versus L	1.44245	-4.13682	0.99488	0.07633
PD/f^2 versus m	1.42164	-2.0 492	0.99612	0.18492
SP versus L	2.61388	0.50935	0.99854	0.08869
N versus L	1.57004	3.73671	0.99531	0.10262
N versus m	0.55749	5.31118	0.99998	0.01071
emf versus L	3.6471	0.87525	0.99227	0.24471
emf versus m	0.70315	2.96107	0.99905	1.0453
f versus L upper	−3.73843	4.77188	0.99787	0 10906
f versus L lower	−1.78155	1.17333	0.99613	0.071639

R: correlation coefficient, SD: standard deviation: PD: volume power density, m: proof mass, L: scaling length, f: frequency, SP: specific power, emf: electromotive force

Keratin has been combined with chitosan so as to create a keratin/chitosan triboelectric nanogenerator device for the simultaneous treatment of waste and the generation of energy[26]. The material was prepared by freeze-drying in order to overcome the poor mechanical properties of keratin and to ensure a high area of surface contact. The output voltage was increased by 375%, as compared with that of the material without keratin; yielding 322V under 6N. The power-density was $14.4W/m^2$ and was maintained for more than 8000 cycles.

Hydrogels which possess good mechanical strength and fatigue resistance are important to the development of triboelectric nanogenerators, in which a flexible electrode is the key component. The preparation of such electrodes tends to be complex however, and the mechanical strength is poor. A suitable wood-based hydrogel composite was proposed[27] which was composed of acrylamide and 2-acrylamido-2-methyl-1-propanesulfonic acid hydrogel and delignified wood. It combined the fibre-structure of wood with the softness of hydrogel, and offered good mechanical properties and fatigue resistance. It wa incorporated into a triboelectric nanogenerator which exhibited a high power-generation ability. The triboelectric nanogenerator had an open-circuit voltage of 60V and was sensitive to mechanical stimuli, thus making it applicable to wearable motion-monitoring.

Liquid metals have been widely used as flexible electrodes for triboelectric nanogenerators, but there remains a problem with metal droplets, due to mismatched surface tension. Co-polymer surface-modified liquid-metal nanocapsules with a core–shell structure were synthesized[28] by using a 2-step method involving chelation reaction and *in situ* free-radical polymerization. The as-prepared nanocapsules could be incorporated into various polymers to create flexible nanocomposites. Introduction of the nanocapsules could improve the charge-storage and electrical output of various nanocomposites. An optimized polydimethylsiloxane/nanocapsule-based triboelectric nanogenerator increased the power-density by 28.3 times.

Polymer-based triboelectric nanogenerators for flexible wearable electronics suffer from a lack of flame-retardancy. An ultra-thin highly-flexible aramid-nanofibre/$Ti_3C_2T_x$/Ni nanochain composite paper was prepared[29] by using vacuum-assisted filtration and freeze-drying. Due to a synergy between the aramid and the MXene, the composite paper exhibited good mechanical properties and a Joule heating ability. The material offered an open-circuit voltage of 55.6V, a short-circuit current of 0.62μA and a transferred charge of 25μC. It could be used to create a self-powering wearable device with an instantaneous power of 15.6μW with an optimum external resistance of 10MΩ.

Fabric-based triboelectric nanogenerators offer a good output, flexibility and wearability but the fabric structure can create gaps which accumulate contaminants. A superhydrophobic fabric-based triboelectric nanogenerator woven with superhydrophobic electroconductive bacterial cellulose fibre was developed[30] in order to counter the problem. The material exhibited a maximum open-circuit voltage of 266.0V, a short-circuit current of 5.9μA and an output power of 489.7μW.

Triboelectric fabrics are important for wearable power-supply, but imparting both good electrical properties and good stretchability is difficult. Such a fabric was developed[31] by preparing a porous composite with polyethyleneimine/silica-coated titanium dioxide modulated carboxymethyl cellulose/waterborne polyurethane as the tribomaterial and polyethyleneimine/carbon-nanotube conductive nanomaterial as the electrode. By adjusting the mass ratio of the components, the electrical performance of the optimized triboelectric fabric could be improved by 111% and could offer a power-density of 1.62W/m^2. The improvement was attributed to the synergistic effect of an enhanced internal polarization and a double charge-transfer mechanism which originated from amino groups. The triboelectric fabric offered an elongation of 245% and an air-permeability of 105.1mm/s.

Possible damage to triboelectric materials affects their practical application. A flexible triboelecric nanogenerator was therefore developed[32] which used self-healing hydrogel and fluorinated ethylene propylene film as triboelectric materials for mechanical energy-harvesting. The hydrogel exhibited good flexibility, transparency and self-healing abilities. It also offered good mechanical properties, with no plastic deformation or damage after stretching by 200%. The output was up to 33.0V and 3μA for a contact frequency of 0.40Hz and a pressure of 2.9N, respectively. This could charge a 0.22μF capacitor to 24.3V within 300s. The voltage retention-rate after self-healing was up to 88.0%. The hydrogel-based nanogenerator could function as a wearable pressure-sensor and offered a sensitivity of 105.9mV/N or 1.73nA/N at a contact frequency of 0.40Hz.

Thin and soft triboelectric nanogenerators are important for power-management in wearable electronics. A simple layout was proposed[33] which comprised 2 soft silicone layers and a graphene-coated fabric layer. This exhibited great flexibility, permeability and durability. It also offered open-circuit voltage and short-current outputs of up to 213.75V and 3.11μA under a constant frequency and stress of 3Hz and 5.6kPa, respectively. The nanogenerators tolerated more than 1000 cycles of bending, stretching and twisting while maintaining those electrical outputs following deformation.

Woody biomass is an abundant renewable resource and aerogels for triboelectric devices have been made from poplar biomass via dissolution and regeneration, by using lithium bromide solution as the solvent. In order to improve structural homogeneity, ball-milling of raw woody biomass before dissolution and ultrasonication following dissolution were used[34]. These treatments modified the porous structure and mechanical properties of the aerogels and led to a marked increase in the triboelectric behaviour. Removal of most of the lignin from the aerogels increased the triboelectric output by 5 times as compared to that of woody biomass aerogel which was made without ball-milling, ultrasonication or lignin reduction.

The combination of a triboelectric nanogenerator and natural wood is a sustainable approach, due to the latter's biodegradability and abundance. On the other hand, cellulose-based nanogenerators can suffer from brittleness, low crystallinity and low surface charge density. A simple method was proposed[35] for modifying natural wood so as to impart good flexibility and tensile properties. Following pressing, the crystallinity was also almost tripled. Cellulose on the wood surface was cationically modified with 3-chloro-2-hydroxypropyl trimethylammonium chloride via solution-immersion. Following modification, the surface potential was almost doubled as compared with that of unmodified material. Density functional theory was used to calculate the absorption energy between cellulose molecules, and the feasibility of chemical modification. The existence of greater differences between two tribo-layers in terms of energy level produces a high flow of electrostatic charge. Modified pressed-wood based triboelectric nanogenerators could generate a peak current of 9.74µA, a voltage of 335V and a transferred-charge density of 71.45µC/m^2 via contact electrification.

Triboelectric nanogenerators convert mechanical energy into electrical energy via tribo-electrification and electrostatic induction, but a low surface charge-density and easy wear are barriers to their application. The inclusion of hexagonal boron nitride nanosheet and polyvinylidene difluoride composite triboelectric material[36] boosted the surface charge density to 711µC/m^2, together with the friction coefficient. The thermal conductivity of a 2wt% composite membrane decreased by 28.6%, and increased by 22.2% with respect to the matrix, respectively. The optimized nanogenerator offered a peak voltage of 434V, a current of 53µA and a power of 4.84mW.

Liquid-solid based triboelectric nanogenerators can be used[37] for rain-drop energy-harvesting, and hydrophobicity of the triboelectric layer is essential for output-increases. The poor mechanical properties of hydrophobic triboelectric layers greatly limit their application. A stretchable hydrophobic triboelectric nanogenerator having a good output and durability was created by strongly bonding PTFE micro-particles to a flexible substrate. A synergistic increase in charge-transfer between droplet and triboelectric layers increased the transfer-charge density to about 4.74 x 10^{-3}C/m^{2}l. The open-circuit voltage and short-circuit current were more than 7 times and 6 times higher, respectively, than those for nanogenerators without PTFE modification. At 500% stretching, the device exhibited a less-than 20% decrease in its output.

The flexible gels which are used in triboelectric nanogenerators continue to suffer from poor mechanical properties and inferior long-term stability. Multifunctional electronic skins have been manufactured[38] from oriented 2-dimensional nanosheets of eutectogel by using a 1-step curing method. These had a transparency of 96%, an elasticity of 1613% and a toughness of 17.22MPa. A triboelectric nanogenerator which was based on this material offered an open-circuit voltage of 224V, a short-circuit current of 7.1μA and a short-circuit charge of 640nC.

A personal thermal management device was developed[39] which had the self-powering ability to generate heat via triboelectricity. Composites were prepared which comprised vertically-aligned silver-tipped $Ni_xCo_{1-x}Se$ nanowire arrays which had been synthesized on the surface of woven Kevlar fibre sheets and reduced graphene oxide, dispersed in polydimethylsiloxane. The $Ag/Ni_xCo_{1-x}Se$ with reduced graphene oxide generated effective Joule heating (79C at 2.1V) in the composites The woven Kevlar fibre, $Ag/Ni_xCo_{1-x}Se$ and polydimethylsiloxane composite offered a higher (98.1%) infrared-reflectivity and better (54.8%) thermal insulation than did a woven Kevlar fibre and polydimethylsiloxane composite. A woven Kevlar fibre, $Ag/Ni_xCo_{1-x}Se$, polydimethylsiloxane and reduced graphene oxide composite had an impact resistance and tensile strength that were 152.2% and 92.1% higher, respectively, than those of woven Kevlar fibre and polydimethylsiloxane composite. A maximum output power-density of $1.1mW/cm^2$ at a frequency of 5Hz confirmed that there occurred sufficient mechanical energy-harvesting of the composites to permit self-heating.

A biocompatible polyacrylamide/calcium-alginate/ethanol double-network hydrogel hydrogel) was prepared[40] by free-radical polymerization and solvent-exchange. The synergistic effect of solvent replacement and the dissipation-energy of the double-network imparted excellent mechanical properties. The hydrogels offered a stretchability greater than 1500%, a fracture energy of $1917J/m^2$ and an ionic conductivity of 1.53S/m. A wearable strain-sensor which was based upon the material had a strain-sensing ability with an operating range of 0 to 250%, a sensitivity of 1.63 and a rapid response-time. A triboelectric nanogenerator for mechanical energy-harvesting was created by using the double-network hydrogel as the electrode. This exhibited an electrical output of up to 236.8V.

A passive-cooling method involved[41] integrating super-hydrophilic titanium dioxide solvent-free nanofluids into cellulose acetate fibrous membranes in order to achieve efficient energy-harvesting. The phase transformation of TiO_2 nanofluids around the temperature range of 32 to 38C regulated the surface temperature of TiO_2 nanofluid and cellulose acetate fibrous membranes. When compared with plain cellulose acetate fabric, the TiO_2-added fabric offered a solar reflectance of 93.43%, an infra-red emissivity of 85%, a water-vapour transport rate of $12.6g/m^2h$ and a cooling-effect of 4 to 8C. The TiO_2 nanofluid markedly increased the ultra-violet resistance, the mechanical properties and the dielectric properties. It improved the triboelectric output performance, leading to and output voltage of 66V, a current of 2.7μA and a power-density of $82mW/m^2$.

Table 2. Mechanical properties of various films

Film	Tensile Strength(MPa)	Fracture Strain(%)	Toughness(MJ/m^3)	E(GPa)
CNF	186.7	6.74	8.0	6.4
VDE	86.0	4.4	2.3	2.9
VDE/CNF	211.6	3.85	4.2	8.6

CNF: cellulose nanofibre, VDE: vanillin-derived epoxy

An energy-harvesting textile was constructed[42] by using natural silk inspired hierarchical structural designs which provided possibilities for optimising the mechanical and triboelectric behaviours. The resultant materials offered a tensile strength of 237MPa and a toughness of 4.5MJ/m^3, plus a power-output of 3.5mW/m^2 and a 99% conductivity which was retained after 2300000 cyclic deformations without any great change in appearance.

Cellulose nanofibres can suffer from high hydrophilicity, flammability and poor ultra-violet light-blocking. A key factor here is the etherification reaction which favours a good compatibility between cellulose nanofibre film and vanillin-derived epoxy resin (table 2), which creates[43] a cross-linked network structure with an adhesion strength of 16.12MPa, a tensile strength of 211.6MPa and a Young's modulus of 8.6GPa. A thin coating of vanillin-derived epoxy resin on cellulose nanofibre film imparted a hydrophobicity of about 93.0° and an ultra-violet shielding of some 100%. The coated film also exhibited a higher (213C) thermal stability than did the uncoated film. The coated film was used in a triboelectric nanogenerator which offered an output-power of 18.33μW.

Cellulose-based composites tend to exhibit a poor water-resistance because they are cross-linked by non-covalent bonds. Furan groups were grafted onto sisal microcrystalline cellulose fibre and a thermally reversible cross-linked network was formed together with maleimide-modified $CaCu_3Ti_4O_{12}$. The mechanical properties of the composites were improved[44] due to an improvement in the cross-linked structure. With a 1wt%$CaCu_3Ti_4O_{12}$content, the tensile strength was 5.53MPa; an improvement of 346%. With 5wt%$CaCu_3Ti_4O_{12}$, the short-circuit current, open circuit voltage and transferred charge of associated triboelectric nanogenerators were 3.2μA, 85V and 26nC, respectively. The generator maintained an electrical performance of more than 60% at an ambient humidity of 90%.

Self-healing cross-linked polyurethanes were synthesized[45] by using a pre-polymer technique which used the Diels-Alder reaction between a furan ring and bis-maleimide.

The tensile strength of polyurethane film with 10wt% of bis-maleimide was 54MPa and the shape-recovery rate was 96.99 %. It exhibited an excellent self-healing and recycling behaviour. Surface cracks could be healed within 600s under heat-treatment, and re-shaped by hot-pressing while retaining 93.3% of its mechanical properties. A 2cm x 2cm polyurethane triboelectric nanogenerator could generate a short-circuit current, open-circuit voltage, transferred charge and power-density of 3.6μA, 89V, 29nC and 2.5W/m^2, respectively, at a frequency of 1Hz. The self-healing efficiency of mechanical properties and the triboelectric output performance attained 93.3% and 98%, respectively, following re-processing.

Eutectogels made from deep eutectic solvents are potentially key components in flexible triboelectric nanogenerators due to their ionic conductivity and stretchability. A multiple dynamic hydrogen-bond interaction design was explored[46] in order to prepare eutectogels consisting of itaconic acid and lactic acid. The introduction of the latter provided multiple hydrogen bond interactions for eutectogels together with sufficient mobile charges associated with dissociated cations and anions. This improved the mechanical properties and the ionic conductivity of the eutectogels. The interactions also imparted a self-ability, temperature tolerance and strong interfacial adsorption. By integrating triboelectric materials, an eutectogel-based triboelectric nanogenerator was fabricated which offered a power-density of 2.4W/m^2. The device could provide a stable electrical output in the stretched state at temperatures of -20 to 100C following self-healing.

The ferroelectric polymer, polyvinylidene-fluoride-co-trifluoroethyne, is a flexible material which is useful for energy harvesting. In order to improve its performance, the ceramic, $0.65Pb(Mg_{1/3}Nb_{2/3})O_3$-$0.35PbTiO_3$ which possesses very high piezoelectric properties was used as reinforcement[47]. Reduced graphene oxide was also added as a conductive filler in order to promote charge-conduction. An increase in output voltage of almost 3 times resulted. When used as part of a piezoelectric nanogenerator, the composite offered an output current which was about twice that offered by the plain polymer. The maximum output voltage of a triboelectric nanogenerator was 200V and the maximum current offered by a piezoelectric nanogenerator was 30μA.

A healable and shape-memory dual-functional polymer with improved mechanical properties was integrated into a triboelectric nanogenerator[48]. The resultant device was very robust and the voltage and current outputs of the healed device could recover to their initial values without obvious change. The device thus provided reliable power generation. The structure of the polymer included both poly(1,4-butylene adipate) segments and disulphide bonds. The segments acted as a semicrystalline oligomer for introducing a crystalline phase into the cross-linked network, thus further imparting shape-memory properties. The polymer offered good healability, with a healing ratio under strain of more than 96% and a clearly improved mechanical strength of some 10.5MPa as compared to that of other healable polymers (table 3) used in triboelectric nanogenerators. The shape-fixity ratio and the shape-recovery ratio could attain more than 98% during stretching and bending deformation. Triboelectric nanogenerators which were based upon the polymer exhibited healability of electrical output and shape memory ability.

Table 3. Comparison of healable shape memory and other healable polymers

Material	Strength(MPa)	Healing Ratio(%)
prepared SSP	10.5	96.7
epoxy resin-based polysulphide	2.6	90
disulphide bond based vitrimer	39	100
healable PU-PDMS	0.87	97
self-healable PDMS	120	94
PVA slime	2.6	81

A triboelectric material which comprised a foam-like material with an embedded soft electrode was studied[49]. The foam could be folded and compressed, but immediately recovered when the stress was removed. The stability of the triboelectric foam was tested by using a 0.1842g sample which was stamped 14 times using a 158.6lb tester. The output remained essentially unchanged. When driven by using a polytetrafluoroethylene plate, a 135mm x 135mm x 25mm triboelectric sample offered an output power of up to 5.46mW.

Freeze-drying, combined with gradual thermal-imidization, was used[50] to prepare boron nitride nanosheet-filled polyimide composite foam. The foams exhibited useful properties, including cyclic stability under lengthy periods of compressive loading and unloading processes, together with a relatively low irreversible deformation. Following 10000 cycles at a compression strain of 60%, the total strain-loss of a composite with 12wt% of filler was only 14%; 2/3 of that of the pure polyimide foam. The foam composites exhibited a compression-driven triboelectric output. A higher compressive strain and compression-rate led to stronger electrical signals.

Table 4. Voltage and current outputs of cement-TiO_2 triboelectric nanogenerators

TiO_2(wt%)	Ball-Milling(h)	Voltage(V)	Current (μA)
0	0	32.0	2.1
0.1	0	69.8	4.9
0.2	0	102.8	9.0
0.3	0	61.0	5.6
0.4	0	48.8	5.0
0.5	0	43.5	3.4
0.2	6	54.0	2.7
0.2	9	73.5	3.3
0.2	12	87.2	4.0

A cement-based triboelectric nanogenerator was created[51] by using TiO_2 nanoparticles as an 0.2%wt filler in Portland cement. This nanocomposite could generate a high electrical output (table 4) in vertical contact-separation mode; 3 times higher than that of pristine-cement triboelectric nanogenerators. A maximum power-density of $265 mW/m^2$ was attained. The triboelectric nanogenerators could detect motion forces down to 0.6N. The good properties of the triboelectric nanogenerators were attributed to the improved dielectric behaviour of the nanocomposite. The TiO_2 nanoparticles improved the strength of the cement and increased the compressive strength by some 1.3 times over that of pristine Portland cement. Preparation by mechanical mixing imparted a better energy-conversion efficiency than did ball-milling.

The harvesting and storage of energy by using cement-based conductive composites with carbon-fibre fillers was described[52]. Simultaneous optimisation of the electrical and mechanical properties of the composites identified the best concentration to be 1vol% in order to ensure sufficient long-range electrical conductivity while minimizing any detriment of the mechanical properties. Cement-based conductive triboelectric nanogenerators were developed in order to demonstrate the harvesting of various renewable energies. Capacitors which used cement-based conductive composites as their core element were fabricated.

Table 5. Brunauer-Emmett-Teller surface area of woven carbon-fibre nanorods

Material	Surface Area(m^2/g)
woven carbon fibre	0.784
P@MnSe$_2$	235.942
P@CuSe$_2$	297.326
P@Cu$_{0.25}$Mn$_{0.75}$Se$_2$	364.468
P@Cu$_{0.75}$Mn$_{0.25}$Se$_2$	438.842
P@Cu$_{0.5}$Mn$_{0.5}$Se$_2$	524.617

A flexible superamphiphobic SiO$_2$ and poly(vinylidene difluoride)trifluoroethylene film was developed[53] which could harvest electrostatic energy from water. The principle of the water-triboelectric nanogenerator was based upon the triboelectricity which was generated from contact electrification between water and the superamphiphobic film. Surface modification of the film by fluorine increased the superamphiphobicity and harvested energy. The maximum power-density from the harvesting attained 62.5mW/m^2 and could charge a capacitor. The output peak-to-peak open-circuit voltage and current of the flexible superamphiphobic film could attain 20V and 12μA/cm^2, respectively.

A woven carbon-fibre based multifunctional triboelectric nanogenerator and supercapacitor was described[54] in which energy which was generated by the triboelectric nanogenerator was stored in the supercapacitor. The effectiveness of the woven electrode was increased by growing phosphorus doped Cu-Mn selenide nanowires on its surface (tables 5 to 7). The polyester-based outer surface of the supercapacitor acted as the positive electrode for triboelectric nanogenerator and a polydimethylsiloxane-coated nanowire woven carbon fibre acted as the negative electrode. The integrated triboelectric nanogenerators generated a power of 7.4W/m^2 while the supercapacitor offered energy and power densities of 97.21Wh/kg and 54.25W/kg, respectively. The device exhibited a strength of 583.11MPa, a modulus of 37.52GPa and an impact resistance of 69.91J.

Table 6. Mechanical properties woven carbon-fibre nanorods

Material	Tensile Strength(MPa)	E(GPa)	Impact Energy(J)
woven carbon fibre	234.71	17.58	34.69
P@MnSe$_2$	423.62	23.71	44.02
P@CuSe$_2$	441.58	25.45	47.21
P@Cu$_{0.25}$Mn$_{0.75}$Se$_2$	508.04	32.89	55.52
P@Cu$_{0.75}$Mn$_{0.25}$Se$_2$	539.97	33.64	61.83
P@Cu$_{0.5}$Mn$_{0.5}$Se$_2$	583.11	37.52	69.91

Limited durability arising from friction losses have restricted the development of a solid-solid triboelectric nanogenerator, again offering possibilities for solid-liquid triboelectric nanogenerators. A ferrofluid can be advantageous in solid-liquid triboelectric nanogenerators due to its liquidity and magnetization behaviour. A solid-ferrofluid triboelectric nanogenerator for ultra-low frequency vibration-energy harvesting could harvest swing vibrations[55] by exploiting friction electrification between a polytetrafluoroethylene shell and a ferrofluid. In order to increase the speed of movement of charges, a magnetic field was applied to the generator in order to increase the flow-velocity of the ferrofluid. The volume of ferrofluid and the applied magnetic field-intensity were optimized so as to increase the output of the generator. The results were such that, upon applying swing vibrations of 0.1 to 0.5Hz, a frequency of 0.5Hz yielded a peak-to-peak open-circuit voltage of 0.98V, while the maximum instantaneous current was 1.05nA. Meanwhile, the output power was about 1.03nW and the power-density was 0.0426mW/m^3. The peak output power of the generator parallel array attained 18.2nW with a 700MΩ load resistance and a 1μF capacitor could be charged to 1.5V within 150s.

Carbon nanotubes are useful as flexible electrodes in wearable energy devices due to their electrical conductivity, soft mechanical properties, electrochemical activity and large surface area. On the other hand, their electrical resistance is greater than that of metals, and stretching can lead to deterioration of the electrical behaviour. A stretchable electrode which was based upon laterally-combed carbon nanotube networks was therefore developed[56]. An increased percolation between the combed carbon nanotubes ensured a high electrical conductivity even when mechanically deformed. Additional nickel electroplating and the use of serpentine electrode designs further increased the conductivity and the deformability. The resultant stretchable electrode exhibited a sheet resistance which was comparable to that of metal-film electrodes, and the resistance-change was minimal after being stretched by about 100%. The high conductivity and deformability permitted high stretchable energy-harvesting via a wireless charging coil and a triboelectric generator, plus storage using a lithium-ion battery or a supercapacitor.

Table 7. Electrical properties woven carbon-fibre nanorods

Material	Capacitance(F/g)	Energy-Density(Wh/kg)	Power-Density(W/kg)
woven carbon fibre	0.173	11.03	1.04
P@MnSe$_2$	18.92	56.85	27.29
P@CuSe$_2$	24.56	61.32	31.05
P@Cu$_{0.25}$Mn$_{0.75}$Se$_2$	31.15	72.14	39.47
P@Cu$_{0.75}$Mn$_{0.25}$Se$_2$	39.58	85.69	46.56
P@Cu$_{0.5}$Mn$_{0.5}$Se$_2$	47.34	97.21	54.25

Table 8. Elastic constants used to predict piezoelectric potentials

Material	c_{11}(GPa)	c_{12}(GPa)	c_{13}(GPa)	c_{33}(GPa)	c_{44}(GPa)	c_{66}(GPa)
ZnO	209.7	121.1	105.1	210.9	42.47	44.29
BaTiO$_3$	222	108	111	151	61	134
Pb(Mg,Nb)O$_3$-PbTiO$_3$	160.4	149.6	75.1	120	53.8	28.7

Table 9. Dielectric constants used to predict piezoelectric potentials

Material	Density(kg/m^3)	Resistivity(Ωm)	Dielectric Constant
ZnO	5606	3.5×10^8	7.77
BaTiO$_3$	6012	$>10^{11}$	4400
Pb(Mg,Nb)O$_3$-PbTiO$_3$	8093	$>10^{11}$	7093

Piezoelectric harvesters

Numerical estimates were made[57] of the output power and energy-conversion efficiency of piezoelectric nanostructures such as rectangular nanowires, hexagonal nanowires and 2-dimensional vertical thin films; so-called nanofins. Static analyses were made of the maximum piezoelectric potential that could be produced by a $BaTiO_3$ nanowire, a ZnO nanowire and a ZnO nanofin when they were subjected to a constant external force. Dynamic analyses revealed the power-generation capability of these nanostructures when exposed to ambient vibrations. The ZnO nanowire and nanofin were investigated as being typical nanogenerator components. Their dynamic responses were modeled by using a one degree of freedom system with a series of damping ratios. By combining the transfer-functions of mechanical vibration and piezoelectric charge generation, it was possible to define the output power and efficiency as functions of the vibration frequency and amplitude. The material dependence of a dynamic system was based upon various piezoelectric and ferroelectric systems such as ZnO, $BaTiO_3$ and $(1-x)Pb(Mg_{1/3}Nb_{2/3})O_3$-$xPbTiO_3$. There was a clear relationship between mechanical energy-harvesting capability and the morphology, dimensions and properties of the nanomaterials. Static analyses, based upon the properties of the materials (tables 8 and 9), predicted the maximum piezoelectric potentials that could be produced. A size-dependency analysis suggested optimum sizes for nanowire and nanofin structures. The highest voltage output was found for ZnO nanowires while $BaTiO_3$ and ZnO, $BaTiO_3$ and $(1-x)Pb(Mg_{1/3}Nb_{2/3})O_3$-$xPbTiO_3$ could generate greater output powers. Nanowires were expected to be better candidates than bulk material in mechanical energy-harvesting due to their high strain-tolerance: the maximum strain for ZnO nanowires was 7.7% while the maximum strain for bulk ZnO was 0.2%. The higher flexibility and strain-tolerance of nanostructures could also reduce the risk of fracture or damage to piezoelectric materials under high-frequency vibration conditions. Within the safe acceleration range, the maximum output power-densities of bulk lead zirconate titanate, poly(vinylidene fluoride) and microfiber based composites were 1, 0.5 and $0.2 W/cm^3$, respectively. The calculation showed that ZnO nanowires, nanofins, $BaTiO_3$ and $(1-x)Pb(Mg_{1/3}Nb_{2/3})O_3$-$xPbTiO_3$ nanowires could produce values as high as 1055, 2250, 30000 and $11800 W/cm^3$, respectively, if it were assumed that the entire space was filled with nanowires.

Table 10. Comparative bending properties of piezoelectric materials

Material	Bending Strength(MPa)
KNN-epoxy	106
KNN-GFRP	462
70vol%PZT-30vol%Pt	90
PZT-15vol%Ag composite	129
PZT/ZnO nanowhiskers	120
PZT/ZnO nanowhiskers/0.5wt%Nb_2O_5	128
PZT-5A between Kapton layers	186.6
0-3 polymer composite with 0.1vol%PZT	80
50%PZT-50%PVDF with 0.3% carbon fiber	31.32
KNN-CFRP	850
$BaTiO_3$-CFRP	804
PZT/carbon-black/epoxy composite	65
PZT (PIC151)	64.8
PZT	87
PZT-5A	140.4
PZT-5H	114.8
PZT-4	123.2
PZT-8	127.5
quenched $(Bi_{0.5}Na_{0.5})TiO_3$ ceramics	227
monocrystalline PMN-PT	60.6
monocrystalline PMN-PZT	44.9

KNN: potassium sodium niobate, GFRP: glass-fibre reinforced polymer, PZT: lead zirconium titanate, CFRP: carbon-fibre reinforced polymer

Table 11. Comparative piezoelectric coefficients of particulate composites

Particle	Matrix	Fraction	Units	Coefficient(pC/N)
PZT	epoxy	75	wt%	28
KNN	epoxy/GFRP	20	vol%	3.5
KNN	epoxy	32	vol%	8.9
KNN	epoxy/CFRP	30	vol%	5.8
KNN	epoxy/woven CFRP	30	vol%	4.7
KNN	mullite-fibre/polyamide	20	vol%	3
KNN	PVDF	70	vol%	35
KNN	epoxy	10	vol%	5
KNLN	epoxy	10	vol%	19
KNLN	epoxy	10	vol%	13
KNLN-Z	PVDF	80	wt%	39
BTO	epoxy	32	vol%	5.5
BTO	epoxy	0.5	vol%	1.16
BTO	P(VDF-TrFE)	60	vol%	7-8
BTO-ZnO	epoxy	1-10	vol%	0.118
ZnO	PVDF	0.5	wt%	-32
ZnO	PVDF	0.02	wt%	-9.1
KBT-BA	PVDF	30	vol%	8
$Co_3[Co(CN)_6]_2$	PVDF	5	wt%	37

KNLN: $(K,Na,Li)NbO_3$, KNLN-Z: $(K,Na,Li)NbO_3-ZrO_2$, KBT-BA: $K_{0.5}Bi_{0.5}TiO_3$- $BiAlO_3$, CNF: cellulose nanofibre

The piezoelectric and mechanical properties of potassium sodium niobate composites with epoxy and glass-fibre reinforced polymer were determined[58] by using a small punch and nano-indentation tests. The results were compared with those for competing materials (tables 10 and 11). An analytical solution for a piezoelectric composite thin plate, subjected to bending, was used for the determination of bending properties. Thanks to glass-fibre inclusions, the bending strength was increased by some 4 times, while the Young's modulus in the length direction was approximately doubled. In the thickness direction however the Young's modulus decreased by less than half. Energy-harvesting tests revealed that the output voltage of the glass-fibre composites was greater than that of the epoxy composites. The output voltage was about 2.4V for a compressive stress of 0.2MPa, but the presence of glass fibres decreased the piezoelectric constants. Damped flexural vibration energy-harvesting tests were also performed. The epoxy composites broke during testing, while glass-fibre composites with a load resistance of 10MΩ generated 35nJ of energy.

Fluorinated polyethylene propylene bipolar ferroelectret films having a concentric tunnel structure were prepared by rigid-template based thermoplastic moulding and contact polarization[59]. Two types of energy-harvester were based upon the ferroelectret films. The films exhibited appreciable longitudinal and radial piezoelectric activities, together with good thermal stability. The films were thermally stable at 120C, following a reduction of 35%. A power output of up to 1mW was offered by an energy-harvester working in 33-mode at a resonant frequency of 210Hz, and a mass of 33.4g at an acceleration of 1G. For a device working in 31-mode, a power output of 15μW was obtained at a resonant frequency of 26Hz with a mass of 1.9g.

The cantilever piezoelectric energy-harvester is a common means for harvesting energy in situations involving irregular vibrations. As well as the piezoelectric material, the nature of the substrate layer also plays an important role in increasing energy-harvesting. It was argued[60] that the choice of the substrate material and its mechanical properties depend upon the design of the harvester. Varying the size of the cantilever, the substrate-layer thickness and the type of piezoelectric material showed that the best substrate material permitted the harvesting of more than twice at much energy. A so-called greedy approach to optimisation increased energy-harvesting in about half of the choices considered.

An investigation was made[61] of the energy-harvesting ability of porous piezoelectric polymer films to collect electrical energy from vibrations. The parameters which were examined included the void content, and the dielectric, electrical and mechanical properties of films of poly(ethylene-co-vinyl acetate). There was an increase in the harvested current and a decrease in the Young's modulus with increasing void content. Thermal analysis revealed a decrease in the piezoelectric constant of the porous materials. A mathematical model was able to predict the harvested current as a function of the matrix characteristics, mechanical excitation and percentage of porosity. The output current was directly proportional to that percentage. The harvested power markedly increased with increasing strain or porosity and attained values of up to 0.23, 1.55 and 3.87mW/m^3 for poly(ethylene-co-vinyl acetate) compositions of 0, 37 and 65%, respectively.

Low-pressure plasma techniques were used[62] for the development of piezo and triboelectric hybrid nanogenerators. Plasma-assisted deposition and functionalization were used for the nanoscale design of ZnO polycrystalline shells, the formation of conductive metallic cores in nanowires and the solvent-less surface modification of polymeric coatings and matrices. Perfluorinated chain grafting of polydimethylsiloxane would be a means for increasing hydrophobicity and surface charges while retaining its mechanical properties. It was thus possible to produce efficient Ag/ZnO convoluted piezoelectric nanogenerators which were supported by flexible substrates and embedded in a triboelectric architecture. Factors such as the crystalline texture, ZnO thickness, nanowire aspect-ratio were considered in order to optimize the power output of nanogenerators for energy-harvesting from low-frequency vibrations. The simple 3-layer architecture permitted the harvesting of vibration energy for frequencies of 1 to 50Hz.

A non-linear buckling process was used[63] to convert lithographically-defined 2-dimensional patterns of electrodes and thin films of piezoelectric polymer into 3-dimensional piezoelectric devices. More than 20 different 3-dimensional geometries were created. Such structures were then applied to energy-harvesting using tailored mechanical properties and root-mean-square voltages ranging from 2 to 790mV. Those 3-dimensional geometries having an ultra-low stiffness or asymmetrical layout, offered unique mechanical properties that would be difficult to ensure by using conventional 2-dimensional designs.

Piezoelectric vibrational-energy harvesters tend to comprise a cantilevered beam which consists of a support layer and one or two piezoelectric layers with a tip mass. Such a configuration maximized electromechanical coupling, but the mechanical properties of the piezoelectric material could limit the harvester size and resonant frequency. Studies were made[64] of a new type of piezoelectric energy-harvester in which the mechanical properties and resonant frequency of the cantilever beam were decoupled from the piezoelectric component. Such a so-called base-mounted piezoelectric harvester featured a piezoelectric transducer that was mounted beneath the base of the cantilevered beam resonator. The resonant frequency and beam dimensions were then essentially free parameters. A prototype which consisted of a 1.6mm x 4.9mm x 20.0mm polyurethane beam, a piezoelectric transducer and an 8.36g tip-mass produced an average power of 8.75 and 113μW at 45Hz across a 13.0MΩ load under harmonic base-excitations of constant peak acceleration at 0.25 and 1.0g, respectively. There was an increase in full-width at half-maximum bandwidth, from 1.5 to 5.6Hz, upon using an array of 4 individual harvesters of similar dimensions with a peak power-generation of 10.38µW at 37.6Hz across a 1.934MΩ load for 0.25g peak base-excitation. The harvester behaved as a damped mass-spring system.

An aligned cellulose nanofibre composite was studied[65] which exhibited flexibility, transparency and improved mechanical and piezoelectric properties. The aligned cellulose nanofibre composite was prepared by the electrospinning of cellulose nanofibre and polyvinyl alcohol mixtures, followed by the casting of cellulose nanofibre suspensions. The aligned cellulose nanofibre composite was 85.3% transparent and had an orientation-index of 0.54. Because of alignment, the piezoelectric charge constant of the composite was more than twice that of plain cellulose nanofibre film. The piezoelectric properties of the composite were exploited in a vibrational energy harvester, yielding an output power of 5.43nW. Cellulose nanofibre and polyvinyl alcohol films were further explored[66] as vibrational energy-harvesting materials. Cellulose nanofibres were isolated by using chemical and physical methods and their suspensions were spread onto a flat substrate in order to form films. When such a wet film remained in a 5T superconducting magnet for 168h, there was cellulose nanofibre alignment perpendicular to the magnetic field. To improve further the mechanical properties of the film, stretching was carried out. The suspension was mixed with polyvinyl alcohol in order to improve the film's toughness. Stretching the films produced the highest mechanical properties along the aligned direction. The maximum Young's modulus and tensile strength of 50% stretched cellulose/polyvinyl-alcohol films were 14.9GPa and 170.6MPa, respectively. Vibrational energy-harvesting was possible by exploiting the piezoelectric behaviour of cellulose nanofibres. The cellulose nanofibre plus polyvinyl alcohol films ensured better energy-harvesting than did pure cellulose nanofibre film.

An electro-mechanical vibrational energy-harvester was studied[67] in which a beam was excited by external kinematic periodic forces and was damped by an electrical resistor via a coupled piezoelectric transducer. Non-linearities were introduced by stops which limited transverse displacements of the beam. The interaction between the latter and the stops was modelled as a Winkler elastic foundation. The mechanical properties of the piezoelectric layer were taken into account and the beam was modeled as a composite structure. A geometrically non-linear version of Timoshenko beam theory was assumed, and the equations of motion were derived via the principle of virtual work for large deflections. The power-output and efficiency of the system under harmonic excitation were analysed with regard to the effect of the position of the stops, and their length, upon the dynamics of the beam.

Materials Research Forum LLC
https://doi.org/10.21741/9781644903674

The harvesting of electrical power from the non-linear vibration of an asymmetrical bimorphic piezoelectric plate was studied[68] on the basis of classical plate theory, Von Kármán strain-displacement non-linear relationships in the presence of temperature changes. Two piezoelectric layers, with differing thicknesses, covered the top and bottom layers of a sub-structure. The structure was excited by harmonic transverse forces, and relationships were derived for the voltage and harvested electrical power. The analytical relationship for power was based upon parameters such as the thicknesses of the layers, the load resistance, the frequency of harmonic excitation and the mechanical properties of the structure.

The energy-harvesting efficiency of melt-processed polyamide-11 and its nanocomposites was investigated[69] as a function of filler-type. Nanoclays were used as structural modifiers, and nanocomposites were prepared by using layered clays such as Cloisite 20A and 10A in contents of 1, 2, 4 or 5wt%. The addition of layered silicates improved the mechanical properties. Nanocomposites with Cloisite Na^+-loaded polyamide-11 had the best piezoelectric constant. The latter depended not only upon the crystalline phases but also upon the nature of the filler.

Screen-printed piezoelectric composite energy harvesters were based[70] upon $BaTiO_3$ particles in various concentrations that were deposited by screen-printing. Materials with 5wt%$BaTiO_3$ offered the best piezoelectric properties and this was attributed to an increased interfacial polarisation and to stress concentrations which were caused by the addition of the ceramic particles. The printed piezoelectrics were co-cured with glass-fibre or carbon fibre reinforced composites. The harvesters were subjected to sinusoidal excitations and to industrial vibrations. The composites with 5wt%$BaTiO_3$ exhibited the best energy-harvesting behaviour. An optimised generator offered a maximum power of 2µW and 1.5µW for glass-fibre and carbon-fibre substrates, respectively, under 1g peak-to-peak sinusoidal excitation.

Nanoscale arch structures are popular in vibration detection due to their ability to capture resonance in two independent directions. A study was made[71] of the vibration of a small-scale cantilever curved beam structure made from functionally graded carbon nanotubes and piezoelectric surface layers on a viscoelastic substrate. The piezoelectric layers converted motion into electrical signals, thus harvesting the energy associated with high-frequency motions.

Simple low-cost magnet-free bistable piezoelectric energy-harvesters have been tested[72] in order to evaluate their non-linear dynamics and their harvesting of broadband vibrations. The laser-machined bistable structure consisted of a buckled beam and two supporting beams, together with piezoelectric transducers. The integration of buckled beams and constraining supporting beams permitted a configuration which combined a bistable buckled piezoelectric beam under cantilevered boundary conditions, without requiring external operation. The frequency-range showed that this form of harvester exhibited broadband characteristics when compared with a linear piezoelectric beam of similar configuration. The various vibration-modes and their associated energy-harvesting were analyzed. The possibilities of the harvester were investigated by adjusting parameters such as the width of the supporting beam and its thickness. A non-linear energy-harvesting array, comprising 4 harvesters covering adjacent broadbands, was constructed in order to improve the overall performance. This had an associated broadband-width of 13.7Hz at an acceleration of 0.75g.

A quasi zero-stiffness metastructure has been proposed[73] for the simultaneous purposes of vibration-isolation and energy-harvesting. The proposed device comprised a zero-stiffness support which greatly reduced the natural frequency of the system, plus 4 piezoelectric cantilever beams which localized and harvested vibration-energy. The power-output was examined by varying the associated parameters, and performing static and dynamic experiments. Energy could be harvested in the effective vibration-isolation region and, by adjusting the mechanical properties of the piezoelectric cantilever beam, this zone could be changed. By adjusting the electrical parameters, the maximum power and range could also be modified.

The use of stacked piezoelectric energy-harvesting units in roads was investigated[74] with regard to their size. The latter was optimized on the basis on the mechanical response under given operating conditions and the influence of fatigue resistance and ultimate compressive strength was clarified. The electrical and mechanical properties of stacked piezoelectric harvesting devices were greatly affected by their size, and those with a diameter of 15mm and thickness of 30mm, while comprising 25 to 30 layers, offered better electrical and mechanical properties. A unit with a diameter of 15 to 30mm and comprising 25 layers offered an output voltage and output power of 176V and 192.2mW under working conditions of 3.5kN and 8Hz. The output-voltage change-rate was 17.9% at 20 to 60C, and was 84.1% for humidities ranging from 50 to 100%. Such piezoelectric energy-harvesting units exhibited a satisfactory fatigue resistance. The maximum change-rate of the output voltage and the maximum deformation were 1.2% and 1.8mm following 100000 cycles of unfavorable traffic loadings of 3.5kN and 8Hz, respectively. The ultimate compressive strength of the piezoelectric energy-harvesting unit attained 188MPa.

Table 12. Crystallinity and elastic modulus of nanofibre membranes

Membrane	Crystallinity(%)	E(GPa)
PVDFs	35.14	5.96
PVDF-CoSO₄	52.9	10.10
ZIF67@PVDF	43.9	11.99
PVDF/TBAHP	56.41	6.56
PVDF/TBAHP-CoSO₄	55.11	9.23
ZIF67@PVDF/TBAHP	54.14	15.49

A ZIF-67 functionalized polyvinylidene fluoride and tetrabutylammonium hexafluorophosphate tree-like nanofibre membrane was developed[75] as a piezoelectric nanogenerator for energy harvesting (tables 12 and 13). The β-phase content of the ZIF-67 reached 86.94%, increased the fibre rotation-radius and augmented the electromechanical conversion efficiency. The short-circuit current reached 3.87μA and the output voltage was greater than 9.78V. The material also offered an elastic modulus of up to 1.13GPa and, during a 500s test cycle, its layered structure did not exhibit discernible damage or interlayer fibre separation. The ZIF-67@PVDF/TBAHP-TLNM membranes offered a much better piezoelectric performance than that of pure PVDF. The enhancement of the piezoelectric output was due to the TBAHP-induced tree-like fibre structure having provided a large number of fine branched fibres having diameters of between 5 and 100nm. This resulted in an increase in the electroactive phase and in the crystallinity of the composite, together with a corresponding increase in the piezoelectric constant and Young's modulus. This was attributed to the synergistic effect of ZIF-67 and PVDF.

Many of the most widely-used piezoelectric materials, such as lead zirconate titanate, suffer from brittleness. A highly-compressible piezoelectric nanogenerator was based[76] upon wood-sponge that was obtained by delignification. Due to the increased compressibility of the sponge, the wood nanogenerator (15mm longitudinal x 15mm tangential) could generate an output voltage of up to 0.69V. This was 85 times greater than that generated by the untreated wood, and it remained stable following repeated compression for more than 600 cycles.

Table 13. β-phase content and piezoelectric output of PVDF nanofibre membranes

Membrane	β-Phase(%)	Pressure	Output(V)
PVDF/Cu/MWCNTs	64.7	-	6.7
PVDF/SnO$_2$	91.4	-	49.2
PVDF/BaTiO$_3$	82.39	-	12.3
PVDF/MoS$_2$	70	-	9.4
PVDF/CsPbI$_2$Br	87.8	-	9.4
PVDF/AlO-rGO	90	36kPa	36
PVDF/Fe$_3$O$_4$-GO	80.33	1N	0.0017
PVDF/Ag-BaTiO$_3$	-	1N	3.56
PVDF/KNN/ZnO	94	1kPa	8.31
PVDF/KNN/CNTs	84	1kPa	12
PVDF/G-ZnO	62.36	1N	0.84
PVDF/BaTiO$_3$/graphene	91.1	2Hz	11
PDA@BaTiO$_3$/PVDF	63.0	3N	13
ZIF-67@PVDF/TBAHP-TLNMs	86.76	2N	9.78

A method was proposed[77] for piezoelectric energy-harvesting using poly(vinylidenefluoride-co-trifluoroethylene) impregnated BaTiO$_3$ nanoparticles within a 3-dimensional cellulose scaffold. The construction of a methyl cellulose scaffold led to effective stress-transfer with high mechanical flexibility and a markedly enhanced energy-harvesting output. When the cellulose content was 3wt%, the optimum energy-harvesting behaviour was attained, and this involved a power-density of 42μW/cm^3. That was almost 800% higher than that of conventional flexible piezoelectric composites.

Table 14. Output of piezoelectric energy nanogenerator composites

Composite	$V_{oc}(V)$	$I_{sc}(\mu A)$
PVDF	2.08	0.13
PVDF-TiO₂	5.49	0.23
PVDF-TiO₂-0.5%rGO	6.87	0.57
PVDF-TiO₂-0.75%rGO	8.56	0.61
PVDF-TiO₂-1%rGO	10.2	0.78
PVDF-TiO₂-1.25%rGO	6.6	0.45
PVDF-TiO₂-1.5%rGO	5	0.42

An all-fibre wearable electric power nanogenerator consisted[78] of a polyvinylidene difluoride-NaNbO₃ nanofibre non-woven fabric as an active piezoelectric component, plus an elastic conducting knitted fabric made from segmented polyurethane and silver-coated polyamide multifilament yarns as the top and bottom electrodes. Non-uniform distributions of deformation in a compressed nanogenerator governed the complex operating modes in the piezoelectric nanofibre non-woven fabric. The nanogenerator produced a peak open-circuit voltage of 3.4V and a peak current of 4.4μA during cyclic compression tests at 1Hz at a maximum pressure of 0.2MPa. The all-fibre nanogenerator retained its behaviour after 1000000 compression cycles.

A hybrid poly(vinylidene fluoride) based nanocomposite[79] consisted of a fixed quantity of titanium dioxide and various amounts of reduced graphene oxide (tables 14 and 15). As compared with pure poly(vinylidene fluoride), there was a marked increase in the mechanical properties. The best piezoelectric data were 10.2V and 0.78μA for hybrid nanocomposite which contained 1wt% of reduced graphene oxide. The latter increased interfacial polarization and there were increments of 88% and 104% in the tensile strength and elastic modulus, respectively.

A bendable all-organic monocrystalline material was described[80] which exhibited excellent piezoelectric energy-harvesting properties. The soft piezoelectric crystals had a helical structure, with predominantly weak non-covalent interactions. This permitted the creation of flexible electrical energy-harvesting devices by using a polymer matrix. This yielded an instantaneous peak power-density of about 66μW/cm³, with an energy conversion-efficiency of about 41%.

Table 15. Output voltage of hybrid piezoelectric nanogenerators

Material	Preparation	Voltage
PVDF/rGO/TiO$_2$	solution-cast	10.2
PVDF/TiO$_2$/MoS$_2$	drop-cast	17.4
PVDF/ZnO/MWCNT	solution-cast	1.32
P(VDF-TrFE)/TiO$_2$/ZnO	electrospun	23
PVDF/rGO/BTO	solvent-cast	4.1
PVDF/rGO/MoS$_2$	solvent-cast	2.3
PVDF/(BCT-BZT)/rGO	solution-cast & hot-pressed	4
PVDF/BF33BT/GO	solution-cast	3.9
PVDF/NaNbO$_3$/RGO	solution-cast	2.16
P(VDF-TrFE)-rGO-BaTiO$_3$	spin-coated	8.5

A piezoelectric nanogenerator was based[81] upon multilayer composite fibre. Polyvinylidene fluoride-hexafluoryl propylene composite fibre, doped with MXene was used as a piezoelectric functional layer. A multi-layer design improved the piezoelectric sensitivity and mechanical properties. The multilayer composite fibres exhibited a piezoelectric sensitivity of 10.88V/kPa, such that the output could be described by:

$$Voltage(V) = 10.88P(kPa) - 9.01$$

Under moderate pressure, the piezoelectric nanogenerator generated an open-circuit voltage of 25V. The MPFP multilayer fibres (table 16) also exhibited improved dielectric properties, with a dielectric constant of 2.9 and a dielectric loss of 0.025 at 1kHz. The multilayer structure also improved the mechanical properties of the composite fibre. The elastic modulus of a 3wt% MXene composite fibre decreased from 5.84MPa for a single-layer structure to 4MPa for a 4-layer structure. The latter structure also exhibited a strain elongation of 230%.

Table 16. Piezoelectric outputs of various materials

Material	Piezoelectric Output(V)
BCZT/P(VDF-HFP)	2.5
P(VDF-HFP)/PANI-ZnS	3
PVT-C-PFE 0.35	13
WS2QDs-PVDF-HFP	20
CB/PVDF-HFP	17.6
ZnO@C/PVDF	0.08
3MPFP/TPU/3MPFP/TPU	25

Polyvinylidene fluoride was reinforced[82] with a fixed amount of barium titanate and various concentrations of reduced graphene oxide. The effect of more than 0.45wt% of reduced graphene oxide upon the energy-harvesting behaviour was studied by using solvent casting followed by electrode poling. An increase in the electroactive β-phase from 54 to 73% occurred upon adding reduced graphene oxide. The Young's modulus, tensile strength and hardness were markedly improved. Piezoelectric nanogenerators exhibited an increase in output voltage from 0.98V for pure polyvinylidene fluoride to 4.1V for polyvinylidene fluoride plus barium titanate and reduced graphene oxide when the reduced graphene oxide content was 1.25wt%.

A carbon-fibre based composite was created[83] by integrating piezoelectric poly(3,4-ethylenedioxythiophene)/CuSCN-coated ZnO nanorods into carbon-fibre surfaces. The polymer matrices were the flexible polydimethylsiloxane or rigid epoxy. The polydimethylsiloxane-coated piezoelectric composite could act as an energy-harvester and a self-powered sensor for detecting variations in impact acceleration. The output-voltage increased from 1.4 to 7.6V as the impact acceleration ranged from 0.1 to $0.4m/s^2$. When epoxy was the matrix of the carbon-fibre reinforced device, the voltage ranged from 0.27 to 3.53V when the acceleration varied from 0.1 to $0.4m/s^2$. The inferior output was attributed to the greater stiffness of the matrix.

Electricity can be generated by the piezoelectricity of biaxially-oriented polyethylene terephthalate under mechanical deformation. The material was tested[84] as an energy-conversion and self-powered sensor for wearable electronics. When a pressure-pulse, following pendulum impact, with a maximum stress of 926kPa and an impact-velocity of

2.1m/s was applied, a voltage of 60V was produced; with a short-circuit current and charge density of $15\mu A/cm^2$ and $138nC/m^2$, respectively. Because of the orientation and stress-induced crystallization of its polymer chains, the biaxially-oriented polyethylene terephthalate films have good mechanical properties which are retained, and are beneficial to the durability of sensors. The signals which were derived from pendulum impact could be harvested for up to 80 cycles and for up to 40s, yielding short-circuit voltages of 107 and 95V, respectively.

Table 17. Compressive properties of silicone-rubber composites

TiC(phr)	MoS$_2$(phr)	Compressive Modulus(MPa)
0	0	1.55
2	0	1.68
4	0	1.79
6	0	1.95
8	0	1.79
0	2	1.83
0	4	1.85
0	6	2.02
0	8	1.8

Energy-harvesting via piezoelectricity is possible by using rubber composites. These can be created[85] by adding titanium carbide and molybdenum disulfide, reinforcing and conductive fillers, to a silicone rubber matrix (tables 17 and 18). The composites were fabricated by solution-mixing. The TiC and MoS$_2$ acted as reinforcement, thus improving the mechanical properties. Typical properties can typically be a compressive modulus of 1.55MPa for a control sample, which is increased to 1.95MPa by adding 6 parts per hundred of TiC or to 2.02MPa by adding 6 parts per hundred MoS$_2$. The stretchability was 133% for a control sample, and was increased to 153% upon adding 6 parts per hundred of TiC and to 165% by 6 parts per hundred of MoS$_2$. At 30% strain, the output voltage was 3.5mV for 6 parts per hundred of TiC and 6.7mV for 6 parts per hundred of

MoS_2. The predicted values agreed well with experiment up to 6 parts per hundred and then deviated. The deviation was attributed to partial aggregation of the filler particles.

Table 18. Tensile properties of silicone-rubber composites

TiC(phr)	MoS_2(phr)	E(MPa)	Tensile Strength(MPa)	Fracture Strain(%)
0	0	0.56	0.54	133
2	0	0.66	0.61	125
4	0	0.71	0.78	165
6	0	0.65	0.82	153
8	0	0.61	0.61	123
0	2	0.53	0.52	139
0	4	0.68	0.74	144
0	6	0.66	0.85	165
0	8	0.68	0.69	129

Mechanical energy-transfer between a source and an active material in piezoelectric energy-harvesters was investigated[86]. Analytical predictions were used to optimise mechanical energy transfer into a piezoelectric material. These were validated by tailoring the stiffness of a beam to piezocomposites by experiment. In one case, the mechanical properties of freeze-cast piezocomposites (table 19) were controlled by adjusting the volume fraction of piezoceramic and epoxy filler phases. This made it promising to tailor the mechanical properties of piezocomposites so as to maximize mechanical energy-transfer into a material via stiffness-matching.

Carbon microsphere-based rubber composites have been applied[87] to stretchable piezoelectric energy-harvesting devices. The rubber matrix was room-temperature vulcanized silicone rubber, and the carbon microspheres acted as reinforcement and conductive material. The compressive modulus increased with increasing carbon microsphere content from 1 to 5 parts per hundred (figure 1). The increase was attributed to the formation of filler networks within the rubber matrix. Such networks led to interaction between the particles and between the particles and the polymer chains of the matrix. The compressive modulus was 2.1MPa for a control sample and became as high as 3.3MPa upon adding 5 parts per hundred of carbon microspheres. The tensile strength (figure 2) increased up to 4 parts per hundred of filler and then decreased. This was attributed to saturation of the mechanical strength. The Shore A hardness was 26 for a control sample, and this was increased to 40 upon adding 5 parts per hundred of the microspheres. The electrical properties were also markedly affected, with the electrochemical resistance being 22.05 for a control sample and decreasing to $15.91k\Omega$ upon adding 5 parts per hundred of the carbon. The carbon- and oxygen-containing functional groups of the carbon microspheres acted as catalysts which improved the interfacial interaction of the composites so as improve the mechanical properties. When the materials were used in a piezoelectric energy-harvesting generator, the output voltage was about 1.25V, with a durability of 500000 cycles.

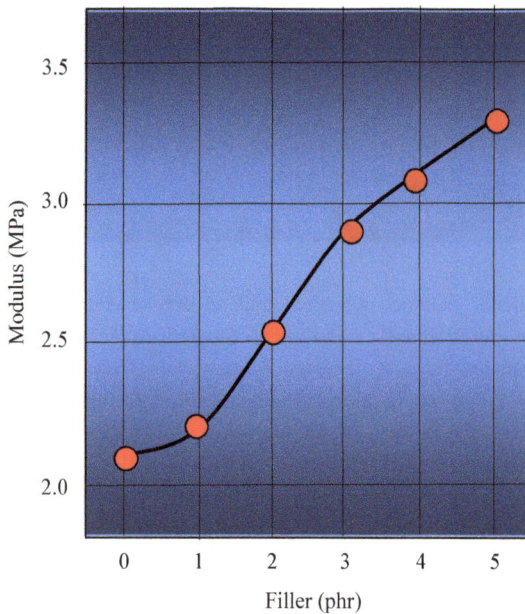

Figure 1. Compressive modulus of rubber composites as a function of carbon microsphere content

Table 19. Properties of freeze-cast lead zieconium titanate composites

Property	PZT-1	PZT-2	NCE55
piezoelectric charge coefficient (pC/N)	580	577	670
piezoelectric voltage coefficient (Vm/N)	0.0282	0.0244	0.019
relative permittivity	2320	2672	5000
PZT volume fraction	0.58	0.73	1.00
stiffness (GPa)	24	39	60

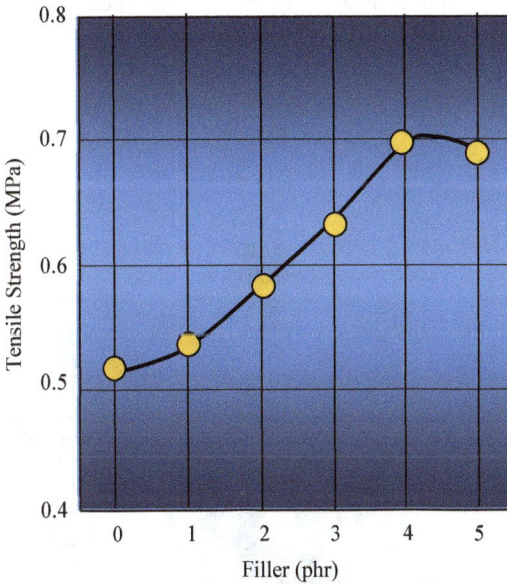

*Figure 2. Tensile strength of rubber composites as a
function of carbon microsphere content*

Table 20. Mechanical properties of polystyrene-BiFeO₃ composites

Filler(wt%)	E(GPa)	Tensile Strength(MPa)	Fracture Elongation(%)
1	0.994	13.820	2.089
2.5	1.065	12.290	2.034
5	0.864	10.535	2.244
10	0.781	11.153	1.913
15	1.555	17.646	1.533
20	1.336	17.127	1.800

The effect of the concentration of $BiFeO_3$ sub-micrometric particles upon the piezoelectric and structural characteristics of composites with a polystyrene matrix was studied[88]. The ferrite particles were created by using the reverse co-precipitation method, followed by ultrasonic dispersion in a solvent to break up any agglomerates which were formed during sintering. This produced particles having a size of about 200nm. Concentrations of $BiFeO_3$ which were higher than 10wt% led to a disturbance in the structure of the polystyrene.

Table 21. Slopes of current-voltage plots and resistance of polystyrene-BiFeO₃ composites

BiFeO₃(%)	Slope(pA/V)	Resistance(GΩ)
1	0.4371	2288
2.5	1.580	633.1
5	0.4356	2296
10	1.839	543.8
15	9.41	106.2
20	1878	0.5326

A hindered organization of side-groups by sub-micrometric filler, which occurred during solvent evaporation, strongly affected the mechanical properties of the composite; markedly increasing the Young's modulus and the tensile strength but decreasing the elongation.

Table 22. Voltage generated by piezoelectric generators in an air-stream

BiFeO$_3$(%)	Air-Stream Pressure(bar)	Condition	Voltage(V)
1	4.86	polarised	1.90
1	4.86	unpolarised	1.15
1	11.54	polarised	4.35
1	11.54	unpolarised	4.1
2.5	4.86	polarised	2.20
2.5	4.86	unpolarised	1.37
2.5	11.54	polarised	4.87
2.5	11.54	unpolarised	4.86
5	4.86	polarised	2.40
5	4.86	unpolarised	1.52
5	11.54	polarised	5.07
5	11.54	unpolarised	4.83
10	4.86	polarised	1.85
10	4.86	unpolarised	1.55
10	11.54	polarised	5.53
10	11.54	unpolarised	4.34
15	4.86	polarised	1.34
15	4.86	unpolarised	1.33
15	11.54	polarised	4.22
15	11.54	unpolarised	2.41
20	4.86	polarised	0.94
20	4.86	unpolarised	0.81
20	11.54	polarised	2.59
20	11.54	unpolarised	0.76

This was attributed to the hindrance of side-group interactions by $BiFeO_3$ particles. A decrease in plasticity could limit the use of the material in a piezoelectric nanogenerator because of the increased risk of fracture. Samples having a weight fraction of $BiFeO_3$ of up to 10% had a similar tensile strength, Young's modulus and elongation (table 20).

Table 23. Piezoelectric coefficients of polystyrene-BiFeO₃ composites

$BiFeO_3$(%)	Condition	Piezoelectric Coefficient(pC/N)
1	polarised	44.1
1	unpolarised	22.2
2.5	polarised	48.2
2.5	unpolarised	41.9
5	polarised	45.4
5	unpolarised	34.2
10	polarised	47.3
10	unpolarised	37.1
15	polarised	26.2
15	unpolarised	24.8
20	polarised	23.8
20	unpolarised	14.12

The current-voltage relationships were fitted, and the resistance was deduced from the reciprocal of the slope (table 21). As the $BiFeO_3$ content increased, the resistance decreased and this trend was especially notable when the $BiFeO_3$ content was greater than 10wt%. This was attributed to the higher conductivity of the piezoelectric filler in comparison to that of the matrix material. Thermal depolarization of the nanogenerators could have differing effects, depending upon the content of piezoelectric phase. A voltage of more than 4.8V was generated when a dynamic air-pressure of 11.54bar was applied to a composite which contained 2.5wt%$BiFeO_3$. When the weight fraction was 10% or lower, the decrease in the voltage following thermal depolarization was about 24%. For weight fractions of 15 and 20wt%, the decrease was about 35%. The responses of polarised and unpolarized piezoelectric nanogenerators were compared (tables 22 and 23). There was a marked decrease in output voltage following nanogenerator depolarization. The degree of reduction depended upon the impact force. Only poled nanogenerators which contained a higher fraction particles offered better energy-conversion, especially those with composites that contained 5 and 10wt% of BiFeO₃. At 4.86bar, a poled nanogenerator with 2.5wt% of particles offered a slightly lower output power-density than did the best nanogenerator with 5wt% of BiFeO₃. A comparison was made with other piezoelectric nanogeneratorss (table 24).

Table 24. Performance of thin-film polymer-based piezoelectric nanogenerators

Composite	Filler(wt%)	Voltage$_{open-circuit}$ (V)	Stress
PVDF-BaTiO₃	1	2(peak-to-peak)	compressive
PVDF-BaTiO₃	5	4(peak-to-peak)	compressive
PVDF-BaTiO₃	7	5(peak-to-peak)	compressive
PVDF-BaTiO₃	10	7.2(peak-to-peak)	compressive
PDMS-BaTiO₃	48	3(peak-to-peak)	compressive
PDMS-BiMgFeCeO₆	5	8(peak-to-peak)	compressive
PDMS-BiMgFeCeO₆	10	17(peak-to-peak)	compressive
PDMS-BiMgFeCeO₆	15	7(peak-to-peak)	compressive
PDMS-BiMgFeCeO₆	20	5(peak-to-peak)	compressive
PU-SbSI, poled	20	0.8(peak-to-peak)	pressing (0.37N, 1Hz)
PU-SbSI, poled	20	0.15(peak-to-peak)	pressing (0.37N, 3Hz)
PU-SbSI, poled	20	0.7(peak-to-peak)	pressing (0.37N, 7Hz)
PU-SbSI, unpoled	20	0.1(peak-to-peak)	pressing (0.37N, 3Hz)
PDMS-BiFeO₃, poled	20	0.8(peak-to-peak)	compressive (10kPa)
PDMS-BiFeO₃, poled	30	1.5(peak-to-peak)	compressive (10kPa)
PDMS-BiFeO₃, poled	40	3.6(peak-to-peak)	compressive (10kPa)
Epoxy-BiFeO₃, unpoled	10	7.3(maximum)	air-stream (17.03bar)

Epoxy-BiFeO$_3$, unpoled	20	6.5(maximum)	air-stream (17.03bar)
Epoxy-BiFeO$_3$, unpoled	30	6.1(maximum)	air-stream (17.03bar)
Epoxy-BiFeO$_3$, unpoled	40	5.7(maximum)	air-stream (17.03bar)
PS-BiFeO$_3$, unpoled	10	0.55(peak-to-peak)	air-stream (17.03bar)
PS-BiFeO$_3$, unpoled	2.5	4.86(maximum)	air-stream (11.54bar)
PS-BiFeO$_3$, unpoled	10	4.34(maximum)	air-stream (11.54bar)
PS-BiFeO$_3$, poled	10	1.47(peak-to-peak)	air-stream (11.54bar)
PS-BiFeO$_3$, poled	2.5	4.87(maximum)	air-stream (11.54bar)
PS-BiFeO$_3$, poled	10	5.53(maximum)	air-stream (11.54bar)

PVDF: poly(vinylidene fluoride), PDMS: poly(dimethylsiloxane), PU: polyurethane, PS: polystyrene

Ultrasound energy-harvesting is another promising field and requires the use of flexible piezoelectric materials which offer a high piezoelectric response in the ultrasonic range. Flexible composites of $0.94[(Na_{0.5}Bi_{0.5})TiO_3]-0.06BaTiO_3$ and polyvinylidene fluoride with filler contents of up to 50vol% (table 25) had a complex connectivity pattern, with fully sintered crystalline powders in the polyvinylidene fluoride matrix[89]. The dielectric constant of the flexible composites increased from 10 for polyvinylidene fluoride to 110 for composite films with 50vol% of crystalline content. The high-frequency piezoelectric constant meanwhile increased from 0.2pC/N to 33pC/N for the same samples. A composite with 50vol% of crystalline content had a figure-of-merit for harvested ultrasound energy of 1.54 x $10^{-12}m^3/J$. This was comparable to the figure-of-merit (1.8 x $10^{-12}m^3/J$) for the piezoelectric ceramic itself.

Lead-free piezoelectric materials of the form, $(1-x)Ba_{0.95}Ca_{0.05}Ti_{0.95}Zr_{0.05}O_3$-$(x)Ba_{0.95}Ca_{0.05}Ti_{0.95}Sn_{0.05}O_3$, have been investigated[90] for energy-harvesting purposes (tables 26 to 29). They were synthesized by high-temperature solid-state for x-values ranging from 0.00 to 1.00. A perovskite structure formed in every case, with no impurity phases. The Ca^{2+}, Zr^{4+} and Sn^{4+} ions were well-dispersed throughout the BaTiO$_3$ lattice. Orthorhombic and tetragonal phases coexisted at room temperature, and there was a steady transition between the phases with increasing x-value.

Table 25. Piezoelectric properties of ceramic/polymer composites

Composite	Filler(vol%)	Piezoelectric Constant(pC/N)
NBT-BT/PVDF	40	13.5
NBT-BT/PVDF	50	33
BFO/PVDF	50	26
PZT/PVDF	48	22
BTO/PVDF	45	22
KNLN/PUc)	50	15
KBT/PVDF	50	20
PZT/PVDF	50	16

KNLN, (K,Na,Li)NbO$_3$, KBT: K$_{0.5}$Bi$_{0.5}$TiO$_3$, PZT: lead zirconium titanate

Table 26. Structural parameters of (1-x)BCZT-(x)BCST

x	Space Group	a(Å)	b(Å)	c(Å)	V(Å^3)	Fraction(%)
0.0	Amm2	3.9905	5.6822	5.6723	128.61	69.1
0.2	Amm2	4.0190	5.6475	5.6525	128.29	66.4
0.4	Amm2	3.9958	5.6671	5.6793	128.60	60.01
0.6	Amm2	3.9978	5.6589	5.7360	129.76	52.2
0.8	Amm2	4.0192	5.6491	5.7105	129.65	45.3
1.0	Amm2	3.9824	5.6567	5.6717	127.76	39
0.0	P4mm	4.0092	4.0092	4.0345	64.84	30.9
0.2	P4mm	4.0088	4.0088	4.0372	64.87	34.6
0.4	P4mm	4.0074	4.0074	4.0371	64.83	39.99
0.6	P4mm	4.0072	4.0072	4.0379	64.83	48
0.8	P4mm	4.0065	4.0065	4.0371	64.80	54.7
1.0	P4mm	4.0055	4.0055	4.0379	64.78	61

The phase-transition temperatures: rhombohedral to orthorhombic, orthorhombic to tetragonal and tetragonal to cubic, shifted to lower temperatures with increasing x-value. Much-improved dielectric and ferroelectric properties were observed, with a dielectric constant of 1900 to 3300 at near to room temperature, and one of 8800 to 12900 near to the Curie temperature. A high electric-field induced strain of 0.12 of 0.175%, a piezoelectric charge coefficient of 296 to 360pC/N, a converse piezoelectric coefficient of 240 to 340pm/V, a planar electromechanical coupling coefficient of a 0.34 to 0.45 and an electrostrictive coefficient of 0.026 to 0.038m^4/C^2 were measured. The composition with x = 0.4 offered the best efficiency with regard to generating electrical energy and thus energy harvesting.

Table 27. Density, grain size, Curie temperature and piezoelectric constant of (1-x)BCZT-(x)BCST

x	ρ(g/cm^3)	grain size(μm)	T$_C$(C)	Piezoelectric Constant(pC/N)
0.0	5.48	26.8	110	296
0.2	5.51	27.6	104	322
0.4	5.53	28.5	98	360
0.6	5.57	29.1	94	350
0.8	5.59	30.1	88	320
1.0	5.62	31.7	80	302

An analytical model, based upon modified shear deformation theory, was used[91] to develop energy-harvesting involving a piezoelectric layer deposited onto functionally-graded beams. The mechanical properties of the beam varied according to a power-law in the beam-thickness direction. Various shear-deformable beam theories were considered, including trigonometric, exponential, hyperbolic and fifth-order shear deformation, that could treat rotary inertia impacts. The harvester was excited at its base, and the response was experienced in the form of translation and rotation. A concentrated mass was placed at the free end of the beam, and Kelvin-Voigt damping or air-damping was assumed to act. The coupled electromechanical governing equations of this functionally-graded harvester were derived by using Hamilton's principle, discretized via Galerkin and diagonalization methods to derive its electrical characteristics. The effects of geometrical parameters, functional-gradient index and concentrated mass upon the voltage and power output were determined with regard to choosing the size and location of the piezoelectric layer.

Table 28. Piezoelectricconstant of BaTiO$_3$-based piezoelectric ceramics

Material	Piezoelectric Constant(pC/N)
4/6BZT/BCT	332
5/5BZT/BCT	500
BCSTZS	514
BZT-BCT-xBST	540
$Ba_{0.88}Ca_{0.12}Ti_{0.94}Sn_{0.06}O_3$	220
$B_{0.95}Co_{0.05}Ti_{0.95}Zr_{0.05}O_3$	296
0.6BCZT-0.4BCST	360
0.4BCZT-0.6BCST	350
$B_{0.95}Co_{0.05}Ti_{0.95}Sn_{0.05}O_3$	302

Coaxial electrospinning was used[92] to prepare polyvinylidene fluoride and polymethyl methacrylate piezoelectric fibre films having a shell-core structure. The fibre films were cut into 2cm x 2cm squares and hot-pressed, every 3 pieces. The product exhibited increased piezoelectric properties, together with a high flexibility. The interaction between CO bonds in the polymethyl methacrylate and the CH_2 in the polyvinylidene fluoride promoted the arrangement of fluorine atoms in the polyvinylidene fluoride. This increased the content of piezoelectric active phase. The coaxial electrospinning encouraged the vertical orientation of hydrogen bonds between the polyvinylidene fluoride and polymethyl methacrylate, thus increasing cross-linking and improving the mechanical properties. When the polyvinylidene fluoride to polymethyl methacrylate layer volume-ratio was 5:1, the piezoelectric coefficient (table 30) of the composite fibre film underwent a 60% increase, from 10pC/N for plain polyvinylidene fluoride to 16pC/N.

Table 29. Energy-harvesting parameters of (1-x)BCZT-(x)BCST

x	Piezoelectric Constant(pC/N)	Figure-of-Merit($10^{-15}m^2/N$)
0.0	296	5200
0.2	322	4667
0.4	360	5496
0.6	350	5260
0.8	320	4088
1.0	302	3108

Polyvinylidene fluoride and trifluoroethylene piezoelectric composites were created[93] by incorporating piezoelectric $ZnSnO_3$ nanoparticles in concentrations of up to 15wt%. The dielectric permittivity of the composite, without nanoparticles, gradually increased with increasing $ZnSnO_3$ content, from 10.98 for the matrix to 13.72 for 15wt% $ZnSnO_3$ loaded material at 1kHz. This was explained in terms of dipolar and space-charge polarization. An increase in the dielectric-loss tangent and losses of the ferroelectric hysteresis loops of samples, upon increasing the filler content, suggested that there was a marked effect of increased space-charge polarization. The nanohardness and elastic modulus increased with increasing filler content, from 43MPa and 1.44GPa for the matrix material to 216MPa and 2.98GPa for a 15wt% content of $ZnSnO_3$. Samples with 10% of $ZnSnO_3$, when used in a nanogenerator, offered an open-circuit voltage of 48V and a power-density of $70\mu W/cm^2$.

Proposed multi-layered ceramics have been based[94] upon a combination of piezoelectric layers of $BaTiO_3$ or $(Ba_{0.85}Ca_{0.15})(Zr_{0.1}Ti_{0.9})O_3$ alternating with mechanically durable ZrO_2 layers. The $BaTiO_3/ZrO_2$ and $(Ba_{0.85}Ca_{0.15})(Zr_{0.1}Ti_{0.9})O_3/ZrO_2$ laminates were electrophoretically deposited. The electrophoretic deposition was followed by conventional sintering or spark plasma sintering. Conventional sintering led to the development of cracks, pores and chemical reaction at the layer interfaces in both types of laminate. Laminates which were sintered by spark plasma sintering were crack-free, with little or no chemical reaction at the interfaces. Nano-indentation tests monitored interlayer porosity evolution and the reaction zone at the interface.

Table 30. Piezoelectric coefficient and dielectric constant of fibre films at 100Hz

Film	Piezoelectric Coefficient(pC/N)	Dielectric Constant
PVDF	10	7.78
PVDF/PMMA-1	16	8.12
PVDF/PMMA-2	12	7.91
PVDF/PMMA-3	8	7.64
PVDF/PMMA-4	6	7.60
PVDF/PMMA-5	6	7.39
PVDF@PMMA	11	7.84

PMMA: poly(methyl methacrylate), PVDF: poly(vinylidene fluoride)

Multiwall carbon nanotubes and partially-reduced graphene oxide were introduced into hybrid composites which were based upon room-temperature vulcanized silicone rubber. The composites were prepared[95] by mixing rubber solution with the nanotube and graphene oxide fillers. This led to better mechanical properties and to reinforcement of the rubber matrix. The partially-reduced graphene oxide was prepared from graphite nanoplatelets by using the modified Hummer method, followed by chemical and thermal reduction. The graphene oxide consisted of less than 5 layers, stacked in a crystalline arrangement. Piezoelectric energy harvesting measurements were performed, and the results indicated higher voltage generation by carbon nanotube containing composites. There was however a tendency to instability and a decline in performance after a few cycles, due to electrode-cracking. On the other hand, hybrid composites offered stable voltage-generation.

It remains a challenge to combine sufficient mechanical strength and high output in composite-type piezoelectric energy-harvesters. Sandwich-structured composite-type harvesters used[96] piezoelectric barium titanate nanoparticles in a polyvinylidene fluoride plus co-trifluoroethylene co-polymer matrix. By modulating the distribution of piezoelectric nanoparticles in the polymer matrix, a maximum open-circuit output of some 60V and a power-density of about $65\mu W/cm^3$ was obtained at a load resistance of $50M\Omega$. A markedly improved energy-harvesting ability was attributed to stress-enforcement in embedded piezoelectric nanoparticles and an optimized piezoelectric response.

Poly(lactic) acid possesses good mechanical and piezoelectric properties and can be 3-dimensionally printed by fusion deposition modelling. A foamed poly(lactic) acid cellular structure was produced[97] and the degree of foaming was varied by varying the extrusion temperature and flow-rate. The maximum surface potential and charge stability of disk samples were obtained by extruding between 230 and 240C at a flow-rate of 53 to 44% and charging on a heated bed at 85C. The cell-morphology and associated mechanical properties were analysed, and the measured piezoelectric coefficient was 212pC/N.

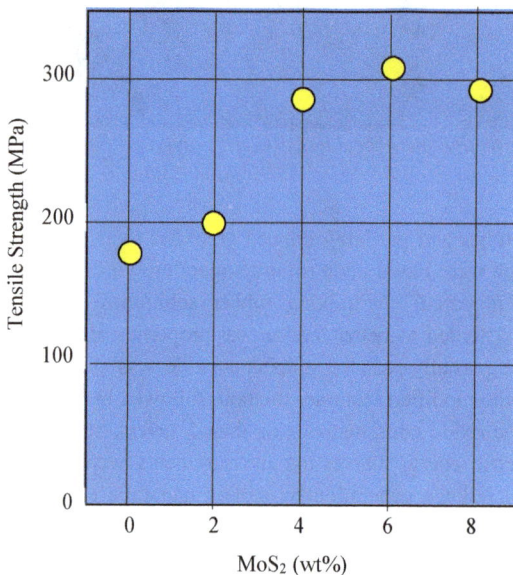

Figure 3. Tensile strength of TOCN-MoS₂ composite films

A piezoelectric hydrogel composite which comprised cross-linked polyvinylidene fluoride and sodium alginate, with calcium copper titanate nanowires as a piezoelectric filler and hydroxyapatite nanoparticles as a mineral filler was developed[98]. Freeze-drying was used to create a piezoelectric hydrogel scaffold. A composite with 0.5% of the titanate and 0.4% of the hydroxyapatite exhibited a greatly improved piezoelectric behaviour and offered the best performance, with an open-circuit output of 7V and a current of 3.5µA. These values were 5.5 and 4.9 times higher than those for the composite hydrogel scaffold itself. The solidity of the hydrogel led to a higher elastic modulus when compared to that of a more viscous medium. The composite offered compressive and tensile strengths of 8.2 and 0.8MPa, respectively, as compared to those of the pure hydrogel: 6.3 and 0.4MPa, respectively. The material was able to charge a 1μF capacitor.

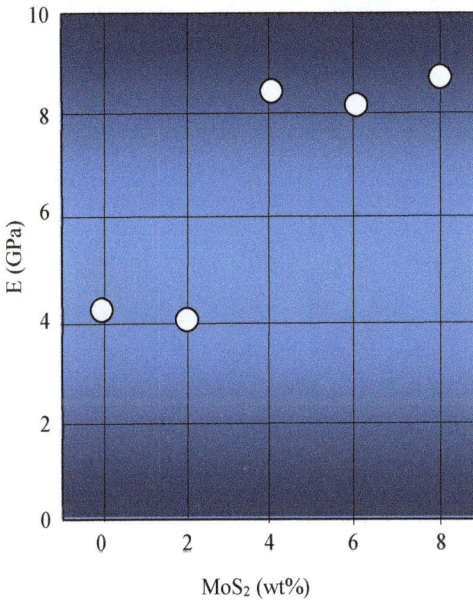

Figure 4. Young's modulus of TOCN-MoS₂ composite films

Graphitic carbon nitride (g-C$_3$N$_4$) can be used for energy-capture by exploiting the strong piezoelectricity which arises from intrinsically non-centrosymmetrical holes. This leads to in-plane piezoelectricity in multilayers or bulk structures. A study of nitride-based piezoelectric nanogenerators revealed[99] that a multi-fold improvement in output could be achieved by modulating precursors via tuning of the intrinsic lattice strain and crystallinity. Output voltages and current densities of up to 1672mV and 141nA/cm^2 could be obtained by using blended urea and melamine precursors.

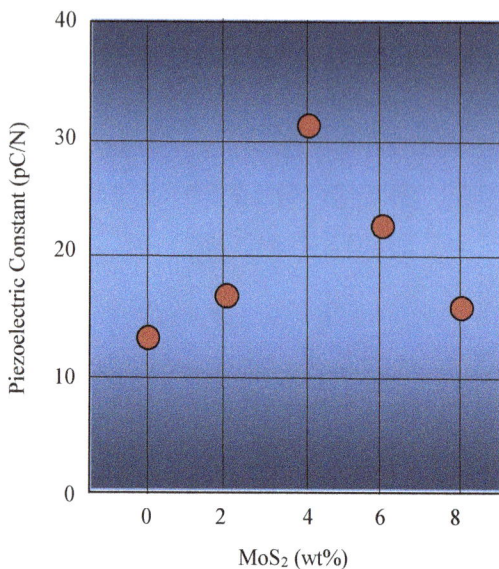

Figure 5. Piezoelectric constant of TOCN-MoS₂ composite films

Biomass-based piezoelectric nanocomposites which comprise piezoelectric ceramic fillers and a natural-polymer matrix are desirable for flexible piezoelectric nanogenerators having high outputs. Composite piezoelectric films have been prepared[100] by incorporating MoS_2 nanosheets and tetragonal $BaTiO_3$ nanoparticles into a 2,2,6,6-tetramethylpiperidine-1-oxyl-oxidized cellulose nanofibril matrix by using a simple method. This ternary composite film exhibited a maximum tensile strength of 335MPa and a maximum longitudinal piezoelectric constant of 45pC/N; much higher than that of other cellulose-based piezoelectric materials. The resultant flexible composite film-based piezoelectric nanogenerator offered a maximum open-circuit voltage of 8.2V and a maximum short-circuit current of 0.48μA. The electricity which was generated could charge a 10μF capacitor up to 3.1V at 110s.

Some 2,2,6,6-tetramethylpiperidine-1-oxyl-oxidized cellulose nanofibril/molybdenum disulfide nanosheet bio-nanocomposite piezoelectric films were prepared[101] by aqueous dispersion. The nanocomposite films exhibited good mechanical properties (figures 3 and 4), with a maximum Young's modulus of 8.2GPa and a tensile strength of 307MPa. When the MoS_2 content was 4wt%, the longitudinal piezoelectric constant (figure 5) of the composite film attained a maximum value of 31pC/N. This was much higher than the 12pC/N of the plain film and other cellulose-based piezoelectric films. Nanogenerators which were made from the MoS_2-containing composite films offered a maximum output voltage of 4.1V and a short-circuit current of 0.21μA. The current which was generated by the nanogenerator was rectified and converted into direct current, and a 10μF capacitor was charged to 1.6V in 120s.

Table 31. Characteristics of room-temperature vulcanised silicone rubber and chemically-reduced graphene oxide composites

Composition	$\Delta G(J/mol)$	$\Delta S(10^2 J/mol)$	Crosslink-Density($10^4 mol/cm^3$)
RTV	-68.04	22.83	4.76
RTV-0.5CRGO	-69.04	23.17	4.64
RTV-1.0CRGO	-63.52	21.32	4.37
RTV-2.0CRGO	-11.74	3.94	0.90

The effect of the shape of piezoelectric beams was investigated[102]. The shapes included triangular, trapezoidal, rectangular, inverted trapezoidal, convex parabolic, concave parabolic and comb-shaped (2 triangular beams with a 4mm-long connector). The analysis was performed for a bimorph piezoelectric beam, and parameters which were considered included the beam length, thickness and width of the piezoelectric layer. The results revealed that, due to the mechanical properties of the beams, the natural frequency of the triangular beam was greatest for all of the considered parameters. As the width of the beam-end increased, the natural frequency decreased. Because the natural frequency was inversely related to the electric power, the inverted trapezoidal beam offered the highest electric power and the triangular beam offered the lowest.

The piezoelectric polymer, polyvinylidene fluoride, possesses suitable mechanical properties for handling large-amplitude motions, has high piezoelectric coefficients and transparency and can be easily integrated into devices. A transparent piezoelectric (23pC/N) energy-harvesting device was created[103] with screen-printed transparent (72%) and conductive (42Ω/sq) electrodes on polyvinylidene fluoride sheets, generating 12μW and 8μW under pressing and bending; corresponding to an energy per cycle of 37nJ and 55nJ, respectively.

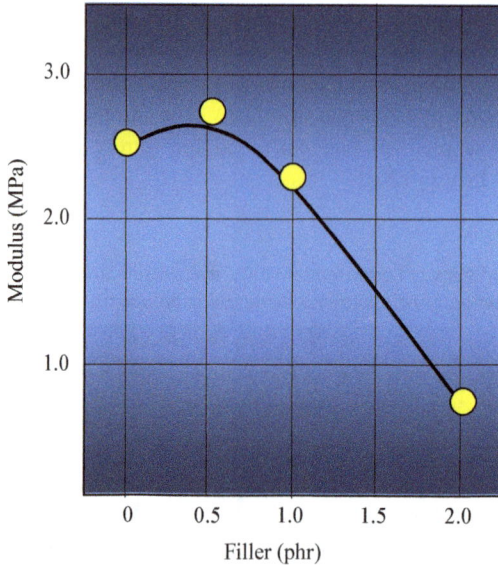

Figure 6. Compressive modulus of room-temperature
vulcanised silicone rubber and chemically-reduced graphene
oxide composites as a function of filler content

Chemically-reduced graphene oxide was synthesized[104] by means of a modified Hummer method. The reduced graphene oxide and room-temperature vulcanized silicone rubber were made into composites with various graphene oxide loadings (table 31). As little as 2 parts per hundred of chemically-reduced graphene oxide in the rubber transformed the previously rigid vulcanized silicone rubber into a soft composite. The compressive modulus (figure 6) fell to 0.6MPa. The fracture strain increased to 100% and the tensile strength (figure 7) fell to 0.1MPa. The electrical conductivity increased as the content of chemically-reduced graphene oxide was increased. These soft composites, with higher electrical properties, were useful for soft piezoelectric energy-harvesting. The output voltages were stable for room-temperature vulcanized silicone rubber, but increased with increasing cycle number for vulcanized silicone rubber plus reduced graphene oxide composites.

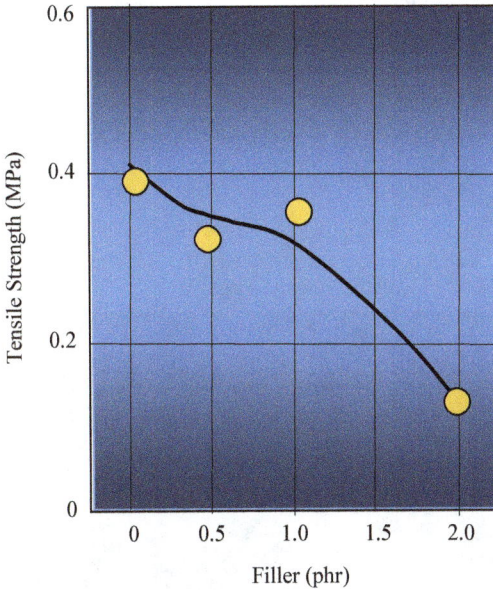

*Figure 7. Tensile strength of room-temperature vulcanised
silicone rubber and chemically-reduced graphene oxide
composites as a function of filler content*

Electrospinning has advantages for the production of poly(vinylidene fluoride)-based nanofibrous structures without requiring secondary processing in order to control the piezoelectric properties. Two methods were used[105] to control the dimensions of electrospun poly(vinylidene fluoride-trifluoroethylene) in order to produce nanofibres with diameters ranging from 1000 to less than 100nm. The size-reduction led to an increase in the effective content of piezoelectric phase and the degree of crystallinity. The changes in crystal structure in turn led to a 2-fold increase in the piezoelectric constant. The size-reduction also increased the Young's modulus of the nanofibres by up to about 80 times. The increases in piezoelectric constant and Young's modulus improved the piezoelectric behaviour, leading to an exponential increase in the electrical output of nanofibre mats when the fibre diameter was reduced from 860nm to 90nm.

Nanowire pellets of antimony seleno-iodide, SbSeI, have been fabricated for use in piezoelectric nanogenerators (table 32) for the conversion of mechanical energy into electrical energy[106]. The nanowires were prepared sonochemically and then compressed at 120MPa. A maximum open-circuit voltage of 384.7mV corresponded to a maximum surface power-density of $14.1nW/cm^2$ and a volume power-density of $0.380\mu W/cm^3$ under periodic striking excitation using a force of 17.8N and a resonant frequency of 70Hz. It could be used as a self-powered sensor for the detection of dynamic changes with frequencies of up to 200Hz.

Table 32. Comparison of the open-circuit voltage of various nanogenerators

Nanogenerator	Voltage(V)
ZnO nanowire array	0.0065
ZnO nanostrands	0.15
ZnO nanosheet network	0.1
PAN/SbSI composite	0.400
BaTiO$_3$ nanotube arrays	0.7
PZT textile	0.4
SnS$_2$ nanosheets	0.012
monolayer MoS$_2$ flakes	0.052
PVDF	0.013
PVDF/BaTiO$_3$ composite	0.528
LiNbO$_3$/PDMS composite	0.46
SbSeI pellet	0.3847

Carbon-black and few-layer graphene were added to co-polymers of poly(vinylidene fluoride and hexafluoropropylene in order to produce hybrid composite films via solution-casting[107]. By adjusting the contents of carbon-black and few-layer graphene, the relative fraction of β-phase could be markedly increased, leading to a high piezoelectricity of the hybrid composite films. The maximum output voltage and the harvesting power-density of the films were 181% and 329%, respectively, as compared to those of pristine films.

An investigation was made[108] of cantilever-type shape-memory alloy and piezoelectric laminated composite beams which were subjected to a tip-load and temperature variations. Such a beam consisted of two piezoelectric layers of identical thickness which were bonded to a superelastic shape-memory alloy core that exhibited an asymmetrical tension-compression behaviour. Either the temperature of the beam or the tip-load was varied, while the other parameter remained fixed. In each case, the load resulted in deformation of the shape-memory alloy core and its bonded piezoelectric layers, leading to the generation of electric charge. The deformation of the shape-memory alloy was modeled on the basis of constitutive relationships which accounted for any intrinsic tension–compression asymmetry. Geometrical and force equilibrium considerations were used to identify the sequence in which solid-phase structures developed within the superelastic core during a complete loading–unloading cycle. Temperature-dependent moment and shear force equations were then used to investigate the electrical and mechanical properties of the beam.

Table 33. Mechanical properties of piezoelectric composites

POSS(wt%)	Hardness(GPa)	E(GPa)	Contact Stiffness(10^5N/m)
0	0.045	1.52	1.035
3	0.051	1.78	1.187
5	0.059	1.91	1.268
8	0.060	1.97	1.330

POSS: polyhedral oligomeric silsesquioxane

Poly(vinylidene difluoride) polymers offer the highest piezoelectric coefficients and are therefore used as generators, although their relatively low performance can limit their application. Piezoelectric poly(vinylidene-trifluoroethylene) hosts, comprising various amounts of polyhedral oligomeric silsesquioxane, were prepared[109] by using a low-temperature solvent-evaporation method. The polyhedral oligomeric silsesquioxane exhibited good compatibility with poly(vinylidene difluoride) and did not affect crystalline-phase formation in the matrix, leading to good piezoelectric properties. Piezo- and triboelectric nanogenerators exhibited average output voltages and current-densities of 3V and 0.5μA/cm² when subjected to a force of 38N over an area of 1cm². Nano-indentation tests showed that the hardness and modulus of samples were increased by

20% and 17% upon making small additions of polyhedral oligomeric silsesquioxane. The piezoelectric nanocomposites consisted of 0, 3, 5 or 8wt% of polyhedral oligomeric silsesquioxane and piezoelectric poly(vinylidenetrifluoroethylene) matrices. Piezoelectric effects could be obtained even when the electric field was very low and the sample was not stretched. When the electric field was greater than 70MV/m, the piezoelectric effect was markedly increased. The piezoelectric effect did not appreciably decrease upon adding polyhedral oligomeric silsesquioxane, especially at 3wt%. The polarization effect could be maintained for long periods by suitable poling. There was no obvious effect of polyhedral oligomeric silsesquioxane upon the coercive field. The hardness, modulus and contact stiffness of samples were increased (table 33) by additions of polyhedral oligomeric silsesquioxane. The increase in the mechanical properties was attributed the presence of rigid Si-O nanocages.

A flexible piezoelectric nanogenerator was based[110] upon a textured hybrid film which comprised poly(vinylidene fluoride) and manganese dioxide nanorods. The combination of MnO_2 and poly(vinylidene fluoride) facilitated the formation of a piezoelectric phase, increased crystallinity and led to a greatly improved piezoelectric output. The addition of MnO_2 markedly improved the mechanical properties while slightly increasing the dielectric constant of the nanocomposites. With the optimum loading of 1.0wt%MnO_2, the piezoelectric coefficient was 38pC/N. An output voltage of 3.2V was generated by the hybrid films while that of poly(vinylidene fluoride) was only 1.7V.

An inorganic-organic hybrid piezoelectric nanogenerator was created[111] by incorporating zinc sulphide nanorods into electrospun poly(vinylidene fluoride) nanofibres. When used as an acoustic energy-harvester, the device operated at a resonance frequency of 86Hz with an acoustic sensitivity of 3V/Pa. It also offered a wind-energy conversion efficiency of 58%.

Near-field electrospinning has been used[112] to prepare ordered poly(vinylidene fluoride) and poly(γ-methyl l-glutamate) composite fibres which exhibited an enhanced piezoelectricity. The preparation process improved the piezoelectric properties of the composites, leading to a better orientation of their dipoles, an ultimate stress of 27.47MPa and a Young's modulus of 2.77GPa. By patterning the piezoelectric composite fibres onto a poly(ethylene terephthalate)-based structure with parallel electrodes, a flexible energy-harvester was created which could capture ambient energy to give a maximum peak voltage of 0.08V, a power of 637.81pW and an energy-conversion efficiency of 3.3%. The electromechanical energy conversion efficiency of this energy-harvester was up to 3 times higher than that of pristine individual poly(vinylidene fluoride) or poly(γ-methyl l-glutamate) energy-harvesters.

Magnetic materials which are endowed with temporary magnetization, permit the mechanical assembly of laser-patterned 2-dimensional magnetic materials into 3-dimensional structures ranging from mesoscale 3-dimensional filaments to arrayed centimetre-scale 3-dimensional membranes[113]. The resultant energy-harvesting systems included 3-dimensional piezoelectric devices for the non-contact conversion of mechanical energy.

A samarium-doping method was used[114] to optimize the properties of $Pb(Zn_{1/3}Nb_{2/3})_{0.20}(Zr_{1/2}Ti_{1/2})_{0.80}O_3$ piezoceramics. A Sm_2O_3 additive could refine the grain size and increase the mechanical properties. The Sm_2O_3 addition provoked the formation of a local structural heterogeneity, together with a reduced domain-size, which boosted the transduction coefficient. A cantilever-beam type piezoelectric energy-harvester was constructed from the samarium-modified material, yielding a power-density of up to $489\mu W/cm^3$ under 1g acceleration. Its ability to charge an electrolytic capacitor was increased by some 91% over that of the undoped material.

Bismuth potassium titanate modified by bismuth nickel titanate was investigated[115], showing that all such ceramics had a single perovskite structure. The latter changed from tetragonal to cubic with increasing content of the modifier. The average grain size decreased and the bulk density and hardness increased with increasing modifier content. The improvement in mechanical properties was related to the change in microstructure and density. Ceramics having a modifier content of 5 to 10mol% possessed good electrical properties. The best dielectric properties were exhibited by ceramic with 10mol% modifier. The highest piezoelectric voltage constant was 0.02392Vm/N, and the off-resonance figure-of-merit for energy harvesting was $6.64pm^2/N$, for a 5mol% composition. This was 5.0 times better than that of pure bismuth potassium titanate. The additive also led to an improvement in the magnetization-field hysteresis loop.

Nanocrystalline $0.2Pb(Zn_{1/3}Nb_{2/3})O_3$-$0.8Pb(Zr_{0.5}Ti_{0.5})O_3$ piezoelectric ceramics, doped with $0.5wt\%MnO_2$, were synthesized[116] by spark plasma sintering. The manganese-doped nanoceramics exhibited good mechanical properties, and also high figures-of-merit. Under an acceleration of $70m/s^2$, a harvester which was made from the material could provide an output power-density as as high as $0.162\mu W/mm^3$.

Additive manufacturing can produce magnetostrictive materials which offer an excellent energy-harvesting capability and provide the required mechanical properties. The parameters of a powder-bed laser-fusion process were optimised[117] so as to produce a magnetostrictive $Fe_{52}Co_{48}$ alloy. A plate of that alloy which had a honeycomb structure was subjected to vibrations and impacts, and the results were compared with those for a fully-dense structure. The honeycomb structure had a lower resonant frequency. In vibration tests, it offered a power-density that was 4.7 times higher than that of the fully dense structure. It was 4.9 times higher in the case of impact tests.

Table 34. Adhesives for a magneto-mechano-electric energy generator

Property	3M DP460	Stycast	Duralco 4461	EPO-TEK
viscosity(cPs)	124	133	111	120
glass transition(C)	-	86	130	65
ShoreD hardness	75-80	96	90	85
operating temperature(C)	23	-40 to 130	<260	<300
E(GPa)	1.7	6.7	3.1	2.5
thickness(μm)	33.23	31.39	18.80	41.39
density(g/cm³)	1.14	2.29	1.1	1.19

Magnetostrictive materials can convert alternating magnetic fields, present in the environment, into mechanical vibrations via magnetostriction. Investigations were made[118] of the magnetoelectric conversion ability of two types of composite component in a direct-current magnetic field alone and in a direct-current and alternating-current coupled magnetic field. When coupled with an alternating-current magnetic field, a direct-current biasing magnetic field could increase magnetization by an alternating-current field in the case of Galfenol alloy, and negate the magnetization in the case of nickel. Composites were made by bonding Galfenol alloy and nickel to a piezoelectric transducer. Under an harmonic magnetic excitation of 3Oe, a direct-current biasing magnetic field of 120Oe could increase the open-circuit voltage of the Galfenol alloy harvester from 0.495 to 10.68V, and the output power from 1.6 to 42μW with a matched external resistance of 50kΩ. In an alternating-current magnetic field of the same amplitude, a direct-current biasing magnetic field increased the open-circuit voltage of a nickel-based harvester from 0.117 to 0.837V and the output power from 2.6 to 23μW with a matched resistance of 1000kΩ.

A simple cantilever-type magneto-mechano-electric energy generator comprises a piezoelectric material laminated onto a magnetostrictive metal plate, having permanent magnets as masses. One problem is the mechano-electric coupling at the interface between the piezoelectric material and magnetostrictive metal layer. This depends greatly upon the mechanical properties of the interfacial adhesive. The effects of 4 types of adhesive (table 34) interfacial layer upon the output power have been investigated[119]. An optimized magneto-mechano-electric energy generator with an adhesive interfacial layer which was 18.8μm in thickness and had an elastic modulus of 3.1GPa led to a maximum output power-density of 0.92mW/cm^2 in the presence of a 10G and 60Hz magnetic field. The generator could endure 10^8 fatigue cycles and exhibited temperature stability at between -30 and 70C.

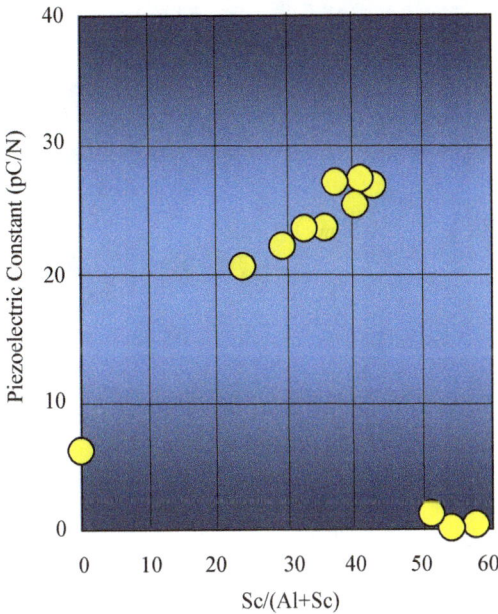

Figure 8 Piezoelectric coefficient and Young's modulus of Al$_x$Sc$_{1-x}$N as a function of the fractional scandium concentration

The deposition of AlN and Al$_x$Sc$_{1-x}$N films by pulse magnetron sputtering was studied[120] with regard to the effect of the doping AlN with scandium upon the piezoelectric (figure 8) and mechanical (figure 9) properties. There was the expected increase in the piezoelectric properties as well as a softening of the material at higher scandium concentrations. At above a threshold concentration of about 40%Sc in the Al$_x$Sc$_{1-x}$N films, there occurred a separation into an aluminium-rich and a scandium-rich wurtzite phase. At scandium concentrations greater than 50%, the films were not piezoelectric and the films consisted mainly of the cubic ScN phase. The films used used for energy-harvesting purposes, and the scandium doping led to an appreciable increase in the energy which was generated. Upon measuring the alternating-current voltage for a base excitation of ±2.5µm, a power of 350µW was found to be generated under optimum conditions, as compared with the 70µW which was found for pure AlN.

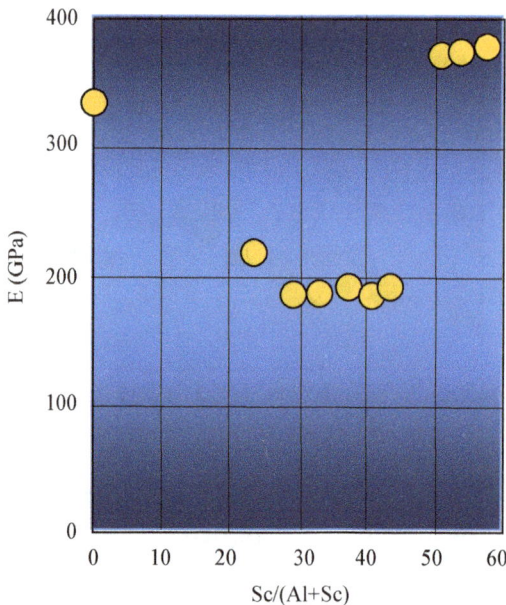

Figure 9. Piezoelectric coefficient and Young's modulus of $Al_xSc_{1-x}N$ as a function of the relative scandium concentration

Piezoelectric zinc oxide nanoparticles were combined with the transparent epoxy photoresist, SU8, in the form of photo-patternable nano-composite films that could be used for energy harvesting[121]. The films were prepared by means of dispersion by ultrasonication, followed by spin-coating and ultra-violet exposure. The films had ZnO concentrations which ranged from 0 to 25wt% and were tested by means of quasi-static and dynamic nano-indentation. The elastic modulus which was measured by quasi-static nano-indentation ranged from 6.2GPa for the pristine SU8 to 8.8GPa for samples with

25wt%ZnO. The hardness ranged from 402MPa to 520MPa for nano-composites having the same range of ZnO contents. The nano-composites exhibited viscoelastic behaviour at frequencies of 10Hz to 201.5Hz.

Composite sandwich-structures were produced[122] which comprised carbon-fibre reinforced polymer face-sheets and specifically designed cores. Three-point bending tests were used to assess the mechanical behaviour of these cellular sandwich-structures. In order to evaluate the energy-harvesting possibilities, the piezoelectric transducer was integrated into the interface between the upper face-sheet and the core and subjected to sinusoidal base excitations and vehicle accelerations. A sandwich which had a conventional honeycomb core exhibited the best mechanical performance during bending tests. With regard to energy-harvesting, a sandwich structure with a re-entrant honeycomb offered a 20% higher root-mean-square voltage output than did sandwich stuctures which had a conventional honeycomb and chiral core. Resistance sweep-tests suggested that the power output from sandwich structures with a re-entrant honeycomb core was twice as high as that from a sandwich structure with a conventional honeycomb and chiral core, under sinusoidal base-excitation.

Vertically aligned nanowire arrays are used for the development of electromechanical transducers. The high aspect-ratio of the nanowires permits large deformations of the nanoscale constituents and this can be exploited in energy-harvesting devices. The vertically aligned nanowires could create a functional gradient that blended the mechanical properties of two dissimilar materials. A piezoelectric energy-harvester was prepared[123] which comprised a functionally-graded nanowire interface rather than a discrete film interface. This harvester generated 7.2 times more power than did film-based devices having the same piezoelectric volume. The improved properties were attributed to the hierarchical structure and the different operation of the piezoelectric constituents.

Porous piezoceramics can exhibit improved energy-harvesting properties, compared with those of dense counterparts, due to a reduction in permittivity which is associated with the porosity. This effect was considered[124] with respect to the influence of the increased mechanical compliance of porous piezoceramics upon the energy-conversion efficiency and the output electrical power (figure 10). Finite-element modelling was used to investigate the effect of porosity upon the energy-harvesting figure-of-merit. The increase in compliance which was due to porosity increased the amount of mechanical energy, which was transferred into the system under stress-driven conditions, and the stress-driven figure-of-merit. This was in spite of a reduction in the electromechanical coupling coefficient. Porosity could be used to tailor the electrical and mechanical properties of piezoceramic harvesters. Two new figures-of-merit were defined which were based upon each stage of piezoelectric harvesting and whether the system was stress-driven or strain-driven.

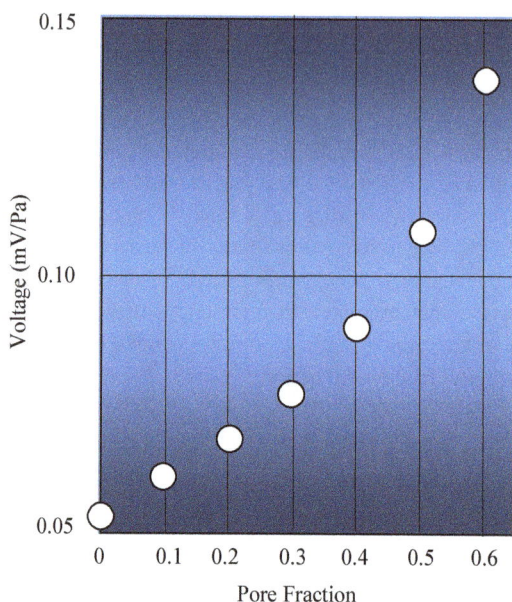

Figure 10. Open-circuit voltage due to a stress applied to porous piezoceramics

Flexible poly(dimethylsiloxane)-BaTiO$_3$ nano-composites were prepared[125] by room-temperature mixing. The effect of BaTiO$_3$ upon the electrical properties was such that the permittivity of the composites was appreciably increased, while the volume resistivity decreased with increasing BaTiO$_3$ concentration. The mechanical properties of the composites were also composition-dependent, with both the tensile strength and fracture-elongation decreasing with increasing BaTiO$_3$ particle concentration. This was attributed to the non-reinforcing nature of the particles. These composites exhibited a good piezoelectric behaviour, with the dielectric properties changing markedly with changes in applied stress. The dielectric properties increased with increasing temperature up to a certain point and then decreased.

A piezoelectric impact energy-harvesting device consisted of 2 piezoelectric beams and a seismic mass. In order to account for the dynamics of the beams, a model was proposed[126] which included the mechanical and piezoelectric properties of the system. The impacts which acted during energy-harvesting were described in terms of a Hertzian contact laws. A transient regime and a steady-state regime were identified and the behaviour of the device was characterized by the steady-state mean electrical power and transient electrical power. Time simulations were to study the effects of seismic mass, beam-length, gap, gliding length and impact location. The latter was an important parameter and could be optimized only via simulation.

The mixed rolling-sliding characteristics and piezoelectric output of sol-gel ZnO films which were deposited onto gold-coated glass substrates and in contact with a rolling-sliding counterface electrode were studied[127]. By exposing the ZnO surfaces to ultra-violet light, their frictional response to rolling-sliding counterfaces could be studied further. The film wetting-behaviour and frictional response were connected. The mechanical properties of the ZnO films were determined by nano-indentation. The piezoelectric response of the films was measured *in situ* by using a reciprocating apparatus under mixed rolling-sliding conditions. The proposed device remained functional for up to 7500 cycles.

The effect of the aspect-ratio of ferroelectric ceramic inclusions upon the piezoelectric performance and hydrostatic parameters of 3-component composites which were based upon relaxor-ferroelectric single crystals was studied[128]. Differences in the micro-geometry of the ceramic/polymer matrix, and the presence of 2 piezo-active components with different piezoelectric and mechanical properties, led to an appreciable dependence of the piezoelectric performance, hydrostatic response and related parameters of the composite upon the aspect-ratio and volume fraction of aligned ceramic inclusions. The piezoelectric performance of $0.67Pb(Mg_{1/3}Nb_{2/3})O_3$-$0.33PbTiO_3$ single crystal and modified $PbTiO_3$ ceramic/polymer composite was valuable for energy-harvesting, in view of a piezoelectric coefficient of 400 to 550mVm/N.

When optimizing a piezocomposite the object is to improve its performance by changing the volume fractions of the constituent materials, the shape of inclusions and the mechanical properties of a polymer matrix. A method which was based upon topology optimization and homogenization was used[129] to design functionally-graded materials for energy-harvesting applications. The effect of the piezoelectric-polarization direction was demonstrated by using a discrete material optimization method which combined gradients and mathematical programming to satisfy a discrete optimization problem. The homogenization method was applied by using a graded finite-element concept which took

Materials Research Forum LLC
https://doi.org/10.21741/9781644903674

account of continuous gradation within the finite elements. It was shown how microscopic stresses could be reduced by combining functionally-graded materials with optimization. The effect of polygonal elements was compared with that of the usual quadrilateral 4-node finite-element meshes. The quadrilaterals involved 1-node connections and were susceptible to checkerboard patterns in topology optimization, so Voronoi diagrams were used to generate irregular polygonal meshes for the purpose of designing piezocomposites.

Waste-heat harvesting

Thermoelectric cement can convert thermal energy to electrical power by exploiting the temperature difference between the road surface and the substrate. During the preparation of large-scale thermoelectric cement, the thermoelectric performance is reduced. Existing thermoelectric cement blocks which offer excellent thermoelectric properties are small, and cannot be used for large-area applications. The thermoelectric properties of large-scale composites which are prepared by pouring suffer from shrinkage-cracking. A thermoelectric cement block which was prepared by dry-pressing could be assembled into a large thermoelectric unit which was more suitable for pavement energy-harvesting[130]. The basic unit was 50mm x 50mm x 20mm, and could be assembled into thermoelectric modules which were 300mm x 300mm x 35 mm. The addition of expanded graphite and metal oxides increased the average conductivity of the blocks to 0.63S/cm, while the average Seebeck coefficient was 20.79μV/K. Use of series connections of the thermoelectric modules can yield a higher energy. Modules which were assembled by using thermoelectric cement-based composites could generate 0.5kWh within 24h from a 1km-long road with a width of 10m. Thermoelectric cement-based composites could also reduce the surface temperature by 1 to 3C.

Wearable thermoelectric materials have the ability to convert body-heat into electricity, but it is difficult to combine good mechanical properties with a high thermopower. A method which was based upon synergistic ion-ion/ion-dipole interactions was used[131] to create ionic thermoelectric ionogels which offered good mechanical properties and a high thermoelectric performance. The as-prepared ionogel offered a stretchability of up to 1160%, a hysteresis of 4.6%, a tensile strength of 5.25MPa and a toughness of 22.09MJ/m^3 together with excellent thermo-mechanical stability and a temperature range of -99 to 250C. The ionogel moreover had an ionic Seebeck coefficient of up to 28.43mV/K, an ionic conductivity of 35.3mS/cm and a power-factor of 2.85mW/mK^2 at a relative humidity of 90%. These all led to an ionic thermoelectric figure-of-merit of 6.9 was attained which the material useful for energy-harvesting.

Organic aerogels possessing good thermoelectric and mechanical properties are candidate materials for use as wearable thermoelectric generators. Most high-performance organic aerogels are made from p-type materials, but it is desirable to develop p-type counterparts. A cellulose nanofibre skeleton was used[132] to form a high-performance n-type organic aerogel which had a Seebeck coefficient of -17µV/K and could be combined with p-type counterparts to create a generator which output 1µW for a temperature-difference of 50K.

The harvesting of low-grade environmental waste heat by using thermoelectric materials is a promising method for dealing with the energy crisis. The mechanical properties of thermoelectric materials are essential to ensuring a long lifespan. Ionic thermoelectric hydrogels, polyacrylamide/carboxymethyl cellulose-xLiCl, possessing good mechanical properties were developed[133]. The hydrogels consisted of a dual network structure, with the LiCl serving as the conductive material. At room temperature, the optimum Seebeck coefficient and ionic conductivity of the hydrogels could attain 2.96mV/K and 36.51mS/cm, respectively. Physical entanglement, hydrogen-bonding and chemical cross-linking of polymer chains within the ionic hydrogels led to a fracture elongation greater than 1300% and a fracture toughness greater than 1700kJ/m^3, together with 95% maintaince of resilience under a strain of 400%. Due to the hydration properties of LiCl, the ionic hydrogels exhibited a good freeze-resistance and the ability to absorb moisture for self-regeneration upon drying. An ionic thermoelectric supercapacitor could generate a thermoelectric voltage of 0.182V under a temperature-difference of 12K, with a power-density of 6.68mW/m^2.

Table 35. Characteristics of Li$_2$ScAuX$_6$ compounds

X	a(Å)	B(GPa)	T$_m$(K)
Br	10.92	28.65	733
I	10.78	21.61	716

Double perovskites are potential materials for harvesting energy from waste heat. A theoretical calculation was made[134] of the electronic, optical and transport properties of Cs$_2$AgAsX$_6$, where X was Cl, Br or I. The unit-cells of these face-centred cubic materials and their lattice constants increased from 10.35 to 11.89Å with increasing ionic radius of the halogen. The enthalpies of formation, of -1.55 to -0.85eV, reflected their structural stability. Band-structure analysis revealed a shift in the band-gap from 1.71 to 0.90eV, from visible to infra-red, upon replacing chlorine with bromine and iodine. The absorption bands had peaks at 3.0eV, 2.5eV and 2.0eV for chlorine, bromine and iodine, respectively. A high value of electrical conductivity, compared with the thermal conductivity, together with a large Seebeck coefficient and figure-of-merit confirmed the suitability of these materials for energy-harvesting devices. The chlorine-based perovskite was ductile while the bromine and iodine perovskites were brittle. Large Debye temperatures and melting points, hardness and low lattice thermal conductivities added to their importance.

Table 36. Transport properties at 300K of Li2ScAuX6 compounds

X	$\sigma(\times10^5/\Omega m)$	$\kappa(W/mK)$	Seebeck Coefficient(mV/K)
Br	1.61	2.07	0.164
I	1.89	1.98	0.140

Stretchable thermoelectric fibre materials are attractive due to their ability to convert human-body waste heat directly into electricity. The preparation of stretchable poly(3,4-ethylenedioxythiophene):poly(styrenesulfonate) fibres with improved properties and continuous p-n alternating structures however remains difficult. Stretchable continuous p-n alternating thermoelectric fibres were prepared[135] by using a simple microfluidic wet-spinning process in which various fibres were used as p- and n-type segments. The power-factors of the p- and n-type fibres were 2.67 and 3.48μW/mK^2, respectively, with a stress of 16 to 19MPa and a strain of 70%.

The gelation of an organic-solvent based electrolyte system which contained a redox couple was applied to thermocell technology and the effect of gelation of the liquid electrolyte, which contained a cobalt bipyridyl redox couple dissolved in 3-methoxypropionitrile upon the performance of thermocells was investigated[136]. Polyvinylidene difluoride and poly(vinylidene fluoride-co-hexafluoropropene) were used for gelation of the electrolyte. The cell performance was improved by optimizing the concentration of the redox couple and the separation between hot and cold electrodes, and the device stability was maintained over 25h of operation.

Table 37. Elastic constants of $Li_2ScAuBr_6$ and Li_2ScAuI_6

Constant	$Li_2ScAuBr_6$	Li_2ScAuI_6
C_{11}	30.57	27.69
C_{12}	13.50	9.28
C_{44}	9.25	7.10
B(GPa)	19.19	15.41
G(GPa)	8.95	7.87
E(GPa)	23.25	20.19
Poisson ratio	0.30	0.28

The properties (tables 35 to 37) of $Li_2ScAu(Br,I)_6$ perovskites were predicted[137] by means of density functional theory calculations. The energy and phonon band-structures were studied with regard to thermodynamic and lattice-vibration stabilities. The band-gaps of 1.96eV and 1.42eV for $Li_2ScAuBr_6$ and Li_2ScAuI_6 corresponded to absorption bands in the visible range. A low lattice thermal conductivity, a Seebeck coefficient and electrical conductivity improved the figure-of-merit.

Table 38. Compressive and bending strengths of cement composites with TX10

TX10(%)	Compressive Strength(MPa)	Bending Strength(MPa)
0	43.38	6.52
0.3	46.43	6.26
0.6	49.65	7.13
0.9	48.32	6.82
1.2	51.57	6.74

Nano-Fe_2O_3/carbon-fibre/cement-based composites have been fabricated[138] by introducing nano-Fe_2O_3, carbon fibres and polyoxyethylene nonyl phenyl ether (TX10) into a cement matrix. The cement consisted of 64.34wt%CaO, 20.76wt%SiO_2, 4.53wt%Al_2O_3, 3.04wt%Fe_2O_3, 1.05%MgO and 2.32wt%SO_3. The carbon fibres had a diameter of 7mm, a length of 3mm, a tensile strength of 3800MPa, an elastic modulus of 220GPa and a density of $1.76g/cm^3$. The addition of the ether promoted the dispersion of the carbon fibres and nano-Fe_2O_3 particles. The composites which contained 0.6wt% of the ether offered better thermoelectric effects, a denser structure, a higher mechanical strength (table 38) and a smaller surface area than did composites with other ether contents. The maximum Seebeck coefficient was 1.1234mV/K (table 39) and the best thermoelectric figure-of-merit was 8.51 x 10^{-5} (85C). The improved thermoelectric properties of these composites were valuable with regard to energy-harvesting.

Table 39. Properties of cement composites with TX10 at 20 to 85C

TX10(%)	S(mV/K)	σ(S/m)	κ(W/mK)
0.3	0.6914	0.139	0.7319
0.6	1.1234	0.139	0.7319
0.9	0.2594	0.139	0.7319
1.2	0.0753	0.139	0.7319

The thermoelectric properties of amorphous and semi-crystalline polyetherimide and single-wall carbon nanotube composites (figure 11) were determined[139]. The nanocomposites were based upon a non-linear polyetherimide with 0.6, 4.4 or 10vol% of the carbon nanotubes, and remained amorphous after adding the carbon. The carbon nanotubes induced crystallization when a linear polyetherimide was used. The electrical conductivity was measured by using a four-probe technique, and higher values were found for the linear polyetherimide films; attaining 20S/m with 10vol% of the carbon nanotubes. Seebeck-coefficient measurements yielded values of 40 and 55μV/K for 0.6 and 4.4vol% contents of carbon nanotubes in the linear polyetherimide nanocomposites. Values of 16 and 47μV/K were found in the case of the non-linear polyetherimide amorphous films. The improvement was attributed to carbon nanotube-induced crystallization in the linear polyetherimide matrix. These nanocomposites were ideal candidates for large-area thermal energy harvesting in the presence of high temperature gradients.

Density functional theory investigations were made[140] of the structural stability, electronic, optical and thermoelectric properties of ternary nitrides, $LiXN_2$, where was X = V or Nb (tables 40 to 42). The calculated formation energies of both nitrides were

Materials Research Forum LLC
https://doi.org/10.21741/9781644903674

negative, confirming their thermodynamic and structural stabilities. Phonon-dispersion spectra revealed positive frequencies for both nitrides, thus indicating dynamic stability. The calculated elastic and mechanical properties also confirmed the mechanical stability of the nitrides. The electronic band-structure and density-of-states spectra confirmed the semiconducting nature of $LiVN_2$ and $LiNbN_2$, with band-gaps of 1.35 and 1.07eV, respectively. The calculated optical properties of the nitrides revealed strong absorption of light in the visible region. The $LiVN_2$ and $LiNbN_2$ had absorption coefficients of 5.10 x 10^5 and 3.80 x 10^5/cm, respectively. The calculated $LiVN_2$ and $LiNbN_2$ optical conductivities were 4.50 x 10^3 and 3.07 x 10^3/Ωm, respectively. The figures-of-merit of $LiVN_2$ and $LiNbN_2$ at room temperature were 2.24 and 3.48, respectively.

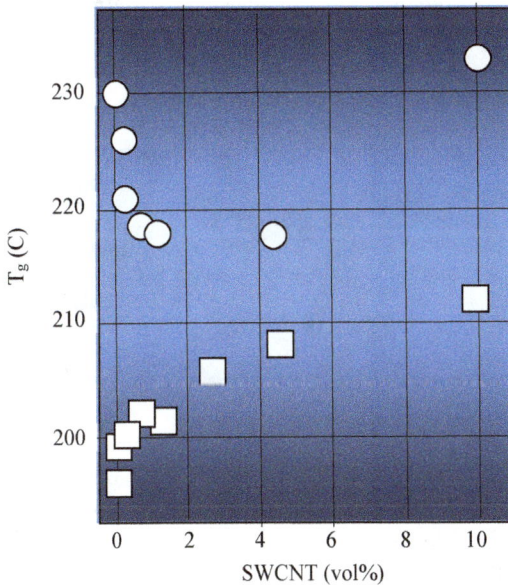

Figure 11. Glass transition temperature of polyetherimide composites as a function of single-walled carbon nanotube content. Circles: 2,30,3,40-biphenyl tetracarboxylic dianhydride -1,4-bis[4-(4-aminophenoxy)phenoxy]benzene, squares: 3,30,4,40-oxydiphthalic dianhydride- 1,4-bis[4-(4-aminophenoxy)phenoxy]benzene

A comprehensive density functional theory and semi-classical Boltzmann transport study has been made[141] of the optical and thermoelectric properties of layer-structured Ba_2XS_4, with X = Zr or Hf (tables 43 to 46), with regard to energy harvesting. There was good agreement between the calculated lattice parameters and experimental data. Both compounds were thermodynamically and mechanically stable, as well as being soft, ductile, machinable and elastically anisotropic. The indirect band-gaps were 1.03eV for Ba_2ZrS_4 and 1.48eV for Ba_2HfS_4. Both compounds exhibited ionic and covalent bonding. The maximum absorption (13.6eV) occurred in the ultra-violet region. The total thermal conductivity increased with temperature, due to increasing electronic thermal conductivity. The total c-axis thermal conductivity at 700K was 4.6W/mK for Ba_2ZrS_4 and 6.1W/mK for Ba_2HfS_4. For p-type Ba_2ZrS_4, the power-factor was about 7mW/mK^2. For Ba_2HfS_4 it was 5.7mW/mK^2. For n-type material, the power-factors were 4 and 3.9mW/mK^2, respectively, along the c-axis at 700K.

Table 40. Calculated structural parameters of ternary nitrides

Nitride	Space Group	a($\AA$)	b($\AA$)	c($\AA$)	V($\AA^3$)	B(GPa)
$LiVN_2$	Pna21	5.442	6.564	5.033	179.768	160.851
$LiNbN_2$	Pnma	6.134	4.244	7.363	191.652	184.084
$ZnSnN_2$	Pna21	5.876	6.762	5.501	262.264	-
$TiFeN_2$	Pnma	4.12	5.78	6.87	163.63	-

Thermoelectrochemical cells are a promising means for the harvesting of waste thermal energy, and the solidification of the redox electrolyte is the key factor. A quasi solid-state electrolyte comprised a ferri/ferrocyanide redox couple within a cellulose matrix[142]. An electrolyte with 5wt% of cellulose offered the optimum balance of mechanical properties, Seebeck coefficients, diffusion coefficients and power-output.

Table 41. Calculated elastic constants of ternary nitrides

Nitride	B(GPa)	G(GPa)	E(GPa)	Poisson Ratio
LiVN$_2$	161.6	51.0	138.6	0.36
LiNbN$_2$	184.8	85.2	221.2	0.30
ZnSnN$_2$	159.4	111.8	271.8	0.22
MgZrN$_2$	220.4	166.6	399.1	0.21

The energy-harvesting possibilities of the thermoelectric effect with regard to concrete surfaces were investigated[143] by incorporating graphite powder (0.5 to 2.5% of cement weight) and steel fibres (1 to 2% of concrete volume). The addition of graphite powder in a 2.5% weight-to-cement ratio led to a maximum upper surface temperature of 50.4C and generated a maximum voltage of 0.27V. The side-surface temperature led to inferior results. It was therefore important to position any device, for converting thermoelectric energy into a current, near to a concrete pavement surface.

Table 42. Calculated thermoelectric parameters of ternary nitrides

Nitride	T(K)	Seebeck Coefficient(V/K)	Hall Coefficient(m^3/C)
LiVN$_2$	300	2.21×10^{-4}	1.91×10^{-8}
LiVN$_2$	1600	1.73×10^{-4}	0.10×10^{-8}
LiNbN$_2$	300	2.33×10^{-4}	1.55×10^{-8}
LiNbN$_2$	1600	1.85×10^{-4}	0.10×10^{-8}
ZnSnN$_2$	300	2.24×10^{-4}	4.00×10^{-8}
ZnSnN$_2$	1600	1.53×10^{-4}	0.06×10^{-8}
ZnMoN$_2$	300	-0.61×10^{-4}	-1.04×10^{-8}
ZnMoN$_2$	1600	-1.11×10^{-4}	-0.29×10^{-8}

Flexible thermoelectric energy-harvesting was achieved[144] via the synthesis of polyaniline-coated Bi$_2$Si$_2$Te$_6$ nanosheet composite films. Their performance was improved by coating the surfaces with polyaniline via solution-mixing. The process involved exfoliation of the Bi$_2$Si$_2$Te$_6$ into nanosheets, followed by coating with camphorsulfonic acid doped polyaniline. Composite films, following 2 coating-cycles, offered a maximum power-factor of about $411\mu W/mK^2$ at 500K, together with flexibility and bending stability; as-prepared polyaniline-coated $Bi_2Si_2Te_6$ nanosheet composite films exhibited great durability following 1000 bending cycles.

Table 43. Calculated lattice parameters of Ba_2ZrS_4 and Ba_2HfS_4

Phase	a(Å)	c(Å)	Volume(Å)
Ba_2ZrS_4	4.783	15.571	356.27
Ba_2ZrS_4	4.715	16.023	356.30
Ba_2ZrS_4	4.785	15.964	365.55
Ba_2HfS_4	4.868	15.624	370.25
Ba_2HfS_4	4.834	15.842	370.20

Density functional theory was used[145] to predict the properties of A_2AuScX_6 double perovskites, where A was Cs or Rb and X was Cl, Br or I. Stability was confirmed by computing the formation energies, binding energies, phonon dispersion and stiffness constants. The band-gaps, absorption coefficients in the visible range and figures-of-merit at 300K were presented (table 47). These compounds were thus potential candidates for thermoelectric devices.

Table 44. Calculated elastic coefficients of
Ba_2ZrS_4 and Ba_2HfS_4

Coefficient	Ba_2ZrS_4	Ba_2HfS_4
C_{11}	116.18	124.51
C_{33}	105.50	106.44
C_{44}	24.36	23.54
C_{66}	22.83	22.18
C_{12}	22.82	22.00
C_{13}	29.41	29.75
G(GPa)	46.68	51.26

A triboelectric nanogenerator and a thermoelectric generator were integrated[146] into an energy-harvesting device for thermal and motional energy in buildings. A fan-shaped rotating triboelectric nanogenerator was compact and easily integrated. The energy-conversion efficiency could attain a maximum of 37.2%. An all-inorganic flexible thermoelectric generator was based upon copper and constantan with an output power-density of $0.73W/m^3$. It retained its original mechanical properties after 10000 bending cycles.

Table 45. Calculated mechanical properties of Ba_2ZrS_4, Ba_2HfS4 and ductile thermoelectric materials

Material	B(GPa)	G(GPa)	E(GPa)	v	B/G	H_V
Ba_2ZrS_4	55.7	30.1	76.6	0.27	1.85	4.13
Ba_2HfS_4	57.6	30.4	77.5	0.28	1.89	3.98
TaCoSn	226.2	99.2	260.0	0.3	2.28	8.22
TiFeTe	148.5	76.7	196.2	0.28	1.93	8.68
NbFeSb	157.9	88.5	223.6	0.26	1.78	9.61

Nanostructures generally improve the performance of thermoelectric materials, but the strength of nano-Sb_2Te_3 based materials largely deteriorates between 573 and 673K. This markedly impairs its use for energy harvesting. The aim is therefore to develop p-type Sb_2Te_3-based materials possessing good mechanical properties and high figures-of-merit over a wide temperature range. Fine grains and nano-twins could be formed[147] simultaneously by reducing the tellurium content and high-energy ball-milling.

Table 46. Calculated Seebeck coefficient of Ba$_2$XS$_4$ at 700K, compared with those of other thermoelectric materials

Material	Carrier	Axis	S(μV/K)
Ba$_2$ZrS$_4$	p	a	123
Ba$_2$ZrS$_4$	p	c	191
Ba$_2$ZrS$_4$	n	a	172
Ba$_2$ZrS$_4$	n	c	170
Ba$_2$HfS$_4$	p	a	151
Ba$_2$HfS$_4$	p	c	185
Ba$_2$HfS$_4$	n	a	168
Ba$_2$HfS$_4$	n	c	167
GeTe	p	-	130
SnSe	p	a	530
SnSe	p	b	518
SnSe	p	c	506
BiCuSeO	p	-	393

This resulted in a lower lattice thermal conductivity and in improved mechanical properties. Doping with copper was used to optimize the carrier and phonon transport behaviours while maintaining good mechanical properties. The figure-of-merit of Sb$_{1.82}$In$_{0.15}$Cu$_{0.03}$Te$_{2.98}$ was increased to 1.06 at 623K and to an average value of 0.76 from 300 to 673K. A p-type segment leg which was composed of this material and Bi$_{0.4}$Sb$_{1.6}$Te$_{3.01}$ was created via 1-step sintering and offered an efficiency of up to 9.2% when the temperature difference was 380K.

Table 47. Properties of A_2AuScX_6 double perovskites

Perovskite	Band-Gap(eV)	Absorption Coefficient(/cm)	Figure-of-Merit
$Cs_2AuScCl_6$	1.88	0.333×10^6	0.44
$Cs_2AuScBr_6$	1.68	0.27×10^6	0.47
Cs_2AuScl_6	1.30	0.213×10^6	0.62
$Rb_2AuScCl_6$	1.93	0.345×10^6	0.47
$Rb_2AuScBr_6$	1.71	0.281×10^6	0.53
Rb_2AuScl_6	1.32	0.218×10^6	0.54

Self-supporting carbon nano-material films have been used[148] to take advantage of the entanglement of carbon nanotubes. By using dimethyl sulfoxide treated poly(3,4-ethylenedioxythiophene) and poly(styrenesulfonate) films, flexible thermoelectric generators were assembled. The effects of the morphology of multi-wall carbon nanotubes, whisker carbon nanotubes and graphene upon the conductivity and mechanical properties were investigated with regard to the thermoelectric properties. The multi-wall carbon nanotube-containing supporting films offered a tensile strength (table 48) of up to 36.23MPa, due to the high-entanglement network-density. Thermoelectric generators which were made from these materials offered high power generation (table 49) and cyclability. The output power-density could be up to $4.6nW/cm^2$ for a temperature-difference of 42.0K, when matched to a suitable load.

Table 48. Output of various thermoelectric generator materials

Material	Tensile Strength(MPa)	Fracture Strain(%)
MWCNT/PEI	36.23	6.08
WSCNT/MWCNT/PEI	32.28	4.50
graphene/MWCNT/PEI	20.45	2.53

Porous graphene aerogel can contain a considerable amount of pure phase-change material in its interior. Volume-shrinkage however seriously limits the weight of working material. But the thermal-energy storage capacity of phase-change material composites depends upon the mass-ratio of pure phase-change material during the phase transition, so graphene aerogel-filled phase-change material composites are suitable for high latent-heat thermal energy storage. Polydimethylsiloxane was introduced[149] into graphene aerogel by spraying. The embedded graphene aerogel exhibited higher mechanical properties and flexibility than those of the pristine aerogel. It effectively reduced the volume-shrinkage and maintained the initial 3-dimensional porous structure while infiltrating pure phase-change material into the interior; leading to an increase in the efficiency of thermoelectric energy-harvesting due to the increase in phase-change material weight.

Table 49. Output of various thermoelectric generators

Material	Type	$\Delta T(K)$	V(mV)	I(μA)	Power(nW)
PDMS/PPy/graphene	foam	10	0.45	-	-
tellurium nanowires/PEDOT	film	45	1.9	1.5	-
graphene/silicone elastomer	film	14.5	0.48	0.13	0.000277
carbon dots/PEDOT:PSS	film	18	4.3	8.6	7.5
PPEK/PEDOT/PEI	fiber	21	1.6	-	0.00161
CNT/carbon nanoparticle	film	5	-	-	120
CNT/biomolecule	yarn	20	4.5	-	-
MWCNT/PEI	film	40	0.23	7.6	1.7
CNT/PEDOT:PSS	film	10	5	-	30
WSCNT/MWCNT/PEI	film	24.1	5.2	2.2	-
PEDOT:PSS	film	42.0	9.9	4.5	11.09

N-MWCNT: nitrogen-doped multi-walled carbon nanotube, NWPU: non-ionic waterborne polyurethane, PDMS: polydimethylsiloxane, PPEK: poly(phthalazinone ether ketone), PPy: polypyrrole, PEDOT: poly(3,-4-ethylenedioxythiophene), PSS: poly(styrenesulfonate, PEI: polyethyleneimine, CNT: carbon nanotubes

Table 50. Calculated elastic constants of cubic SnS

Parameter	Method	Value
C_{11}	local density approximation	133.03(GPa)
C_{11}	generalized gradient approximation	69.66(GPa)
C_{12}	local density approximation	17.46(GPa)
C_{12}	generalized gradient approximation	-3.36(GPa)
C_{44}	local density approximation	22.95(GPa)
C_{44}	generalized gradient approximation	17.31(GPa)
B	local density approximation	55.32(GPa)
B	generalized gradient approximation	20.98(GPa)
β	local density approximation	0.018
β	generalized gradient approximation	0.048
G	local density approximation	33.31(GPa)
G	generalized gradient approximation	23.45(GPa)
E	local density approximation	83.22(GPa)
E	generalized gradient approximation	51.26(GPa)
V	local density approximation	0.25
V	generalized gradient approximation	0.09
μ	local density approximation	33.32
μ	generalized gradient approximation	23.45
λ	local density approximation	33.12
λ	generalized gradient approximation	5.34
B/G	local density approximation	1.66
B/G	generalized gradient approximation	0.89
H_V	local density approximation	6.18(GPa)
H_V	generalized gradient approximation	9.75(GPa)

The cubic π-SnS phase was considered for thermoelectric applications and density functional theory was used[150] to determine its structural and elastic properties (tables 50 and 51). The band structure included a band-gap of 1.073 to 1.372eV. The calculated elastic constants obeyed the Born stability criteria and, on the basis of the Voigt-Reuss-Hill approximation, the bulk modulus, the shear modulus, the Young's modulus, the Poisson's ratio and the Lame coefficients were determined. Surface visualizations of the bulk, shear and Young's moduli indicated that π-SnS is elastically anisotropic. The thermal conductivity of π-SnS could be high due to high a Debye temperature relative to that of α-SnS. The compression and shear wave velocities in 3 directions, [100], [110] and [111], were estimated (table 52).

Graphene aerogel is used as a supporting material for latent-heat thermal energy storage because its 3-dimensional interconnected structure can hold the working material. The mechanical properties were improved[151] in order to reduce the volume-shrinkage of phase-change material composites when the material melts and is incorporated. Polydimethylsiloxane was introduced into the aerogel via solution spray-treatment. Given that the latent heat was proportional to the weight of phase-change material, such a volume-shrinkage could affect the thermoelectric energy-conversion efficiency. The polydimethylsiloxane which was introduced into the graphene aerogel markedly improved the mechanical properties and dimensional stability.

Table 51. Calculated bulk and shear moduli of π-SnS

Parameter	Type	Method	Value(GPa)
B	Voigt	local density approximation	55.32
B	Reuss	local density approximation	55.32
B	Voigt	generalized gradient approximation	55.32
B	Reuss	generalized gradient approximation	20.98
G	Voigt	local density approximation	36.48
G	Reuss	local density approximation	30.13
G	Voigt	generalized gradient approximation	25.00
G	Reuss	generalized gradient approximation	21.92

Single-step water atomization was used to synthesize $Bi_{0.5}Sb_{1.5}Te_3$ powder by copper-doping and consolidation into large-scale bulk samples having a diameter of 50mm and a thickness of some 40mm by using spark plasma sintering[152]. The incorporation of copper boosted the carrier concentration and led to a marked increase in the electrical conductivity, which inhibiting the bipolar thermal conductivity by 73%. The lattice contribution was appreciably reduced by the effective scattering of phonons by the boundaries of fine grains and by point defects. The peak figure-of-merit shifted to the mid-temperature range and went through a maximum value of 1.31 at 425K, with an average value of 1.24 between 300 and 500K for $Bi_{0.5}Sb_{1.5}Te_3Cu_{0.05}$ samples. These values were much greater than those for bare $Bi_{0.5}Sb_{1.5}Te_3$. The maximum compressive strength of large samples was 102MPa. A thermoelectric module which combined $Bi_{0.5}Sb_{1.5}Te_3Cu_{0.05}$ and n-type $Bi_{0.5}Sb_{1.5}Te_3$ offered a cooling performance of -19.4C and a maximum output power of 6.91W for a temperature difference of 200K.

Table 52. Calculated longitudinal and transverse elastic wave velocities in cubic SnS

Direction	Type	Method	Value(km/s)
[100]	longitudinal	local density approximation	4.82
[100]	longitudinal	generalized gradient approximation	3.79
[100]	transverse	local density approximation	2.02
[100]	transverse	generalized gradient approximation	1.89
[110]	longitudinal	local density approximation	4.15
[110]	longitudinal	generalized gradient approximation	3.22
[110]	transverse	local density approximation	2.02
[110]	transverse	generalized gradient approximation	1.89
[111]	longitudinal	local density approximation	2.84
[111]	longitudinal	generalized gradient approximation	3.01
[111]	transverse	local density approximation	3.90
[111]	transverse	generalized gradient approximation	2.74

Energy-Harvesting Materials

Materials Research Forum LLC

Materials Research Foundations **177** (2025)

https://doi.org/10.21741/9781644903674

A modified graphene aerogel which possessed good mechanical properties was prepared[153]. Graphene oxide was used to provide functional groups on the surface via the oxidation of potassium permanganate. The internal structure of the graphene aerogel was cross-linked by using a cysteamine vapour method. The cross-linked graphene/cysteamine aerogel was sufficiently strong to reduce volume shrinkage during the infiltration of phase-change material. The structure could also hold additional pure

Table 53. Elastic constants of π-GeS and π-GeSe

Constant	π-GeS	π-GeSe
C_{11}	55.05	64.29
C_{12}	3.38	4.53
C_{44}	11.89	12.50
B	20.60	24.45
β	0.049	0.041
G	16.31	17.87
E	38.72	43.11
V	0.19	0.21
B/G	1.26	1.37
H_V	5.09	4.96

β: compressibility, V: Poisson ratio

phase-change material within, without any structural damage. A thermoelectric-power generator was created which could produce electrical energy with thermoelectric energy-harvesting efficiencies of 67.92 and 41.72%.

The cubic phases of germanium monochalcogenides (π-GeS, π-GeSe), with their moderate band-gaps and interesting electronic properties, are useful for photovoltaic and thermoelectric applications. The structural and elastic properties (table 53) were investigated[154] by using the ultra-soft pseudopotential technique. Band-structure calculations confirmed that π-GeS and π-GeSe were indirect, with band-gap energies 1.38 and 1.04eV respectively. The calculated elastic constants satisfied Born stability criteria.

The bulk, Young's and shear moduli, the Lame parameters, the Poisson ratio, the Debye temperature and the average sound velocity were determined by using the Voigt-Reuss-Hill approximation. The shear and Young's elastic moduli showed that π-GeS and π-GeSe are anisotropic. The calculated Debye temperatures of π-GeS and π-GeSe were 262.28K and 264.46K, respectively, at 300K. The longitudinal and transverse sound velocities were calculated for the [111], [110] and [100] directions. The germanium sulphide had higher absorption peaks and an optimum band-gap of 1.38eV for use in photovoltaic devices.

A film-based thermopile for energy-harvesting from low-temperature heat sources was described[155]. The device consisted of a silicon frame which comprised alternately-doped polysilicon or poly-SiGe legs that bridged the gap between the hot and cold sides of the frame. It could be created via contact-lithography on silicon wafers. A process was developed for fabricating thermopile structures on membranes of SiO_2/Si_3N_4, Si_3N_4 and low-stress Si_xN_y as well as self-supporting thermopiles. The fabrication of a thermopile structure, without a silicon nitride membrane to eliminate parasitic heat-flow through the membrane, considerably improved the mechanical stability.

Miscellaneous

A study was made of the possibility of exploiting jitter as a renewable power source, while also limiting such jitter[156]. A device was designed in the form of a tuned mass damper-type electromagnetic energy-harvester, combined with a passive vibration-isolator. Also considered was a self-powered system in which disturbances due to jitter were measured by using an accelerometer which was activated by the harvested energy.

The first high-strength fibre-based energy-harvesters were developed[157] by using a simple scalable process. Some light and robust energy-harvesting materials in the form of vertically-aligned ZnO nanowires were grown on the fibre surface of woven aramid fabric. The ZnO nanostructured interface was used as both a functional unit and a reinforcing component. The power-harvesting capability of the hybrid composites was demonstrated via direct vibration of a fabricated cantilever beam. The hybrid composite energy-harvester beam produced an open circuit voltage of about 125mV$_{rms}$, about 0.4V peak-to-peak, when subjected to a cyclic base acceleration of 1G$_{rms}$. The design choices increased the elastic modulus and tensile strength of the composites by 34.3% and 18.4%, respectively, indicating that the integration of ZnO nanowire arrays imparts an energy-harvesting capability and improves the mechanical properties of the composite.

Materials Research Forum LLC
https://doi.org/10.21741/9781644903674

Table 54. Characteristics of the cantilever beam

Parameter	Values
substrate Young's modulus	$97 \times 10^9 \text{N/m}^2$
magneto-electro-elastic material Young's modulus	$2.15 \times 10^{11} \text{N/m}^2$
piezoelectric coefficient	-2.8C/m^2
piezomagnetic coefficient	220N/A
dielectric coefficient	$6.3 \times 10^{-9} \text{C}^2/\text{Nm}^2$
magnetoelectric coefficient	$2750 \times 10^{-12} \text{Ns/VC}$
magnetic permeability	$83.5 \times 10^{-6} \text{Ns}^2/\text{C}^2$
substrate density	8490kg/m^3
magneto-electro-elastic material density	5550kg/m^3
length of the bar	95mm
width of the bar	18mm
thickness of the substrate layer	0.7mm
thickness of the magneto-electro-elastic layer	0.4mm

A parametric study was made[158] of the efficacy of an integrated vibrational energy-harvesting device for wind turbines. A glass-fibre composite device with integrated microfibres was tested by using swept sine vibration under conditions that ranged from -40 to 70C and relative humidities ranging from 10 to 90%. An inverse proportionality was found between the amount of harvested energy and the temperature. The efficiency of the energy-harvesting was governed by the stiffness of the cantilever, which exhibited a viscoelastic temperature dependence. Numerical modal analysis was used to determine the shapes of resonance modes, and their frequencies were temperature-dependent. The humidity had a secondary effect upon energy-harvesting, and there was no appreciable short-term effect upon the structural properties of the samples.

A study was made[159] of the electrostrictive behaviour and vibrational energy-harvesting abilities of conductive polyaniline and polyurethane composites. Polyurethane composites with polyaniline contents of up to 2wt% were prepared by solution-casting. The dielectric and mechanical properties were determined in order to characterize the electrostriction and energy-conversion properties. The polyaniline filler in a polyurethane matrix improved electromechanical coupling due to interfacial charges between the phases. When an external electric field was applied to the electrodes, the current which was generated increased with polyaniline content. The presence of 2wt% of polyaniline in the polyurethane led to the maximum output power in a weak electric field at a 20Hz mechanical frequency.

A magneto-electro-elastic cantilever beam (table 54) was proposed[160] as a device for increasing the harvested electrical power in a vibration-based system. The cantilever beam consisted of a linear homogeneous elastic substrate and 2 magneto-electro-elastic layers with perfect bonding between their interfaces. Any increase in the Young's modulus and density of the substrate layer increased the amount of magnetoelectric harvested power. The positive effect of the Young's modulus of the substrate upon the maximum magneto-electric power was greater at higher densities. The maximum harvested magneto-electric power which was generated with a substrate Young's modulus of 260GPa and a density of 8500kg/m^3 was 3.52 x 10^{-3}Ws^4/m^2.

A silicon-based microelectromechanical sensor having a high sensitivity for certain low frequencies was developed[161] for the purpose of energy-harvesting. It consisted of a disk-shaped mass which was attached to 2 or 3 lead zirconate titanate S-shaped spring flexures. A resonant frequency of less than 11Hz and an output of 7.5mV at an acceleration greater than 0.2g could be achieved.

Dielectric elastomer generators can be used to harvest energy from ambient vibrations. On the basis of the mechanical properties of a circular dielectric elastomer membrane, a dynamic vibro-impact system was developed[162] which converted vibrational energy into electrical energy. The parameters of the membrane were analysed, and used to estimate the output voltage.

A novel method was proposed[163] for the creation of devices which could harvest energy from background mechanical vibrations. The novelties included a non-linear mechanism which controlled the harvester's behavior and the printing method which was used to prepare prototypes. The non-linearity was desirable because vibrational energy tends to be distributed over a low-frequency band and is therefore not easily harvested by a linear resonant device. The ink-jet printing technology which was used was very cheap and thus ideal for rapid prototyping. A study was made of the mechanical properties of a snap-through buckling beam. Power outputs of the order of 100nW were experimentally estimated.

Mechanical non-linearity was shown to be exploitable for use in energy harvesting, where the uncertain nature of the excitation could be problematic. A typical application included harvesting of the energy arising from pedestrian-generated vibrations, which can depend upon individual characteristics and upon the pace. A particular non-linear spring mechanism[164] consisted of a cantilever which wrapped around a curved surface as it deflected. In the case of a free cantilever the force which acts upon the free tip depends linearly upon the tip displacement, whereas the addition of a contact surface having a suitable curvature leads to a non-linear dependence of the tip displacement upon the force acting. This non-linear mechanism offered advantages such as low frictional loss, few moving parts and an ability to withstand excessive loads. By using the non-linear element in a 2 degrees-of-freedom system, it was possible to ensure strongly non-linear energy transfer between the modes of the system. This non-linear behaviour was associated with a marked consistency over 3 very different excitations corresponding to differing walking paces.

A new solution to the problem of cable-vibration mitigation involved the development of an H-bridge based electromagnetic inertial damper which could simulate the control behaviour of an inertial damper and also offer energy-harvesting[165].

At present, most monitoring sensors for railway rolling-stock are powered by external power sources, but the latter might well be replaced by harvesting the vibration-energy generated by trains themselves. A new spherical energy-harvester was designed[166] for use with rubber concrete track-beds. A model analysis identified key factors such as vertical displacement, stress, voltage and electrical energy. It was concluded that, when embedding the harvester as an array within the rubber concrete track bed, the optimum position was located 5mm below the sleeper. The use of 10 energy harvesters within such a 6m-long rubber concrete track-bed generated some millijoules of electrical energy when an 8-carriage train passed over it. The total energy which was captured within 1h amounted to 122.4mJ. Rubber concrete track-beds were better at energy-harvesting than were ordinary ones such that, when the rubber content was 10%, the total electrical energy harvested was 5% higher than that harvested by plain concrete. The mechanical properties were not impaired.

Wearable textiles which offered energy-harvesting, electromagnetic interference shielding and Joule heating were created[167] by depositing cross-linked $Ti_3C_2T_x$, poly(vinyl alcohol) and poly(acrylic acid) onto Hanji (a traditional Korean paper) via vacuum filtration. The textiles exhibited a power-generation which persisted for more than 1h, with a power-density of $102.2\mu W/cm^3$ and an energy-density of $31.0mWh/cm^3$ upon applying $20\mu l$ of NaCl solution. The electromagnetic shielding effectiveness in the 8.2 to 12.4GHz range of up to 437.6dB/mm, with the absorption/reflection ratio reaching 4.5.

Eutectogels are a promising material for flexible electronics but are limited by their poor mechanical properties and ionic conductivity. In order to counter these problems, cationic chitosan quaternary ammonium salt was embedded[168] into the 3-dimensional porous framework of the eutectogel in order to create ion-migration channels. A simple solvent-replacement process increased crystallization of the polyvinyl alcohol matrix and hydrogen-bonding to create composite eutectogels offering high toughness and conductivity. This gel offered mechanical properties of 1.72MPa and 413% together with a conductivity of 0.40S/m. Under an external force, a 3-dimensional porous network with a cationic polysaccharide distribution exhibited an effective piezo-ionic effect. Moderate chitosan quaternary ammonium salt contents greatly increased the piezo-ionic voltage to 270mV; 4.5 times that of the pure eutectogel. The composite eutectogel was used for capacitor energy storage and wearable devices, with 94% capacitance retention and a response-time of 292ms.

Some mechanical energy-harvesters have been based[169] upon stress-voltage coupling in electrochemically alloyed electrodes. They consisted of 2 identical lithium-alloyed silicon electrodes which were separated by electrolyte-soaked polymer membranes. Asymmetrical stresses which were caused by bending generated a chemical potential difference which drove a lithium-ion flux from the compressed electrode to the tensioned electrode so as to generate an electrical current. Removal of the bending reversed the ion flux and the electrical current. A thermodynamic analysis showed that the ideal energy-harvesting efficiency of the device was governed by the Poisson ratio of the electrodes. A thin-film based energy-harvester offerered a generating capacity of 15%.

An anti-freezing hydrogel was synthesized[170] via the 1-step radical polymerization of acrylamide monomer in hydroxyethyl cellulose aqueous solution. After adding LiCl to the hydrogel, it could be cooled to as low as -69C without freezing. The hydroxyethyl cellulose increased the mechanical properties by acting as a cross-linking agent. At a constant elongation of about 150%, an anti-freezing hydrogel-based nanogenerator could harvest human biomechanical energy even in ice and snow. At a frequency of 2.5Hz, a nanogenerator with a 3cm x 3mm area provided an output of 285V and 15.5μA, with an instantaneous peak power density of 626mW/m².

Phase-change materials have a great ability to absorb and release latent heat for thermal-energy storage, and their integration into flexible fibres offers possibilities for wearable systems. A wet-spinning type of core-sheath process was used[171] to create thermochromic phase-change fibres by encapsulating phase-change materials by using octadecane as the core, phase-change materials and polyurethane with thermochromic microcapsules as an encapsulation layer. The resultant materials had an enthalpy of 185J/g together with an elongation of 800%. They also offered an electrical-thermal and solar-thermal conversion and storage efficiency of 90.2% and 91.4%, respectively, when integrated with carbon fibres.

Phase-change materials can improve the efficiency of renewable-energy use by energy-peak shaving and time-shifting. Leakage and inherent rigidity indicate however that an appreciable amount of energy-input is required for the scaling-up of phase-change material applications. High-quality encapsulation requires more complicated processes and a higher energy consumption. A simplified thermally-initiated encapsulation and moulding method has been proposed[172] which involves grafting the phase-change material onto thermosetting polymer backbones so as to construct self-supported phase-change materials having a complex 3-dimensional geometry. By imposing restrictions on liquid materials, an ultra-high precision complex structure can be moulded while mechanical properties are changed by curing so as to impart shape stability. The complete process required no curing-agents or organic solvents. In some cases, a sample could exhibit heat-release for 480s. In simulations, a chip could have an an operating temperature which was 50C lower than that due to natural cooling, thus confirming the potential utility of phase-change materials.

The structural, hydrogen-storage, opto-electronic and mechanical properties of A_2LiCuH_6 double-perovskite hydrides, where A = Be, Mg, Ca or Sr, were determined. A cubic phase and thermal stability were predicted[173] on the basis of formation and cohesive energies, as were hydrogen-storage capabilities. The gravimetric storage capacities were 6.39, 4.83, 3.86 and 2.40%, respectively. The desorption temperatures were 431.8, 586.9, 717.2 and 682.8K, respectively, and the volumetric storage capacities were 21.09, 17.99, 15.68 and $13.69_{gH2}/l$, respectively. The Be_2LiCuH_6 and Mg_2LiCuH_6 were metallic while Ca_2LiCuH_6 and Sr_2LiCuH_6 were semiconductors with indirect band-gaps of 1.03 and 1.58eV. The Ca_2LiCuH_6 and Sr_2LiCuH_6 perovskite hydrides had absorption coefficients of the order of 10^4/cm and were promising candidates for opto-electronic applications.

Photocatalytic water-splitting is a means for hydrogen production, and hydrogen storage is consequently of increasing interest. Perovskite-type compounds such as $GaBaX_3$, where is F, Cl, Br, I or H, are in turn important to this field. All of these compounds were thermodynamically stable, as predicted by formation-energy calculations[174]. A decrease in band-gap was observed when an anionic atom is replaced at the X-site. The optical properties of the compounds were determined from 0 to 30eV and absorption in the visible and ultra-violet range, with high absorption coefficients, ensured they were suitable for harvesting light-energy. These mechanical properties were also favourable and elastically anisotropic, making them ideal for photocatalysis. Suitable band-edge positions were available for redox reactions.

Dielectric elastomer generators are another means for converting motional mechanical energy into electricity, with uniaxial types being particularly suitable due to their simplicity. A silicone-based composite with good dielectric and mechanical properties was developed[175], and a uniaxial generator was designed which had an adjustable aspect-ratio. A 3-dimensional finite-element model revealed the underlying mechanism by analysing the non-uniform deformation of a dielectric elastomer film and its effect upon energy-harvesting. Dielectric elastomer generators with a large aspect-ratio exhibited more uniform and extensive deformation, and this increased the electric-field strength. An excessively high aspect-ratio could however lead to a rapid reduction in the permittivity and an increase in charge-leakage, thus impairing the output. An optimally-designed generator with a moderate aspect-ratio offered a maximum power-conversion efficiency of 40.5% and an energy-density of 54.4mJ/g.

Plane-wave pseudopotential techniques were used[176] to investigate the structural and other properties of Ca-, Sr- and Ca/Sr-doped barium titanate perovskites. A transformation from cubic to tetragonal was observed. The pristine and doped compounds had band-gaps which were suitable for solar-cell applications. Doping increased the density-of-states and thus increased charge mobility. Doping also had effects upon the optical, mechanical and thermal properties. The thermal properties were all suitable for energy-harvesting. All of the doped compounds exhibited mechanical stability. The Poisson and Pugh ratios suggested that doping transformed the material from brittle to ductile, again making it a suitable candidate for flexible solar cells.

Density functional theory was used to analyze[177] the structural, mechanical, electronic and transport properties of Rb_2SeX_6 double perovskite halides, where X was Cl or Br, with the aim of assessing their suitability as energy-harvesting devices. The computed bulk moduli indicated that the compounds possessed mechanical stability. It was deduced that the indirect electronic band-gaps ranged from 2.15 to 2.61eV, so that they fell within the visible region, due to strong orbital coupling of the cations. The replacement of halogen atoms markedly affected the optical characteristics. The data confirmed the potential of the compounds for energy-harvesting.

A sodium potassium niobate nanoparticle-filled epoxy plate was combined[178] with carbon-fibre reinforced polymer electrodes. The composites possessed markedly improved mechanical properties. The carbon-fibre reinforced polymer also contributed to the energy-harvesting output, with a peak-to-peak output voltage of 7.25mV. This was over 600% higher than that for the plate.

A spandex/graphene-cotton/polyurethane yarn was created which comprised a core layer of spandex fibers, a middle layer of graphene/cotton fibres and an outer layer of spindle-knotted polyurethane nanofibres[179]. It exhibited good mechanical properties, superhydrophobicity, strain-sensitivity and fatigue resistance; even under wet conditions. When woven into fabric, it could convert mechanical energy into electrical power, at a maximum voltage of 3.9V and 0.7V for solid-solid and liquid-solid contact, respectively.

Polyurethane has been prepared[180] in 1-step (non-chain extended) and 2-step (chain extended) forms, and both were filled with various concentrations of nanocrystalline cellulose. The mechanical properties greatly affected deformation at the interfaces which occurred during use and the parameters included surface energy and geometry, which affected the contact area. The 1-step polyurethane, with 0.1% cellulose, had a maximum capacitance of 29.20pF, a voltage of 2.04V, a current of 0.43µA and power of 0.89µW. This was a 74.5% increase in the power as compared to that of plain 1-step polyurethane. Improvements were associated with lower concentrations of cellulose nanocrystals, increased hydrogen-bonding and better surface energy. The improvements in output were attributed to improved internal polarization arising from nanocellulose, an increased crystallinity of soft segments and reduced charge-transfer due to amino groups in the chain extender.

A sandwich structure having neck-embedded cavities was used[181] to design a cellular metamaterial which combined sound-absorption, compression/impact resistance and energy-harvesting properties. Resonant cavities could harvest broadband acoustic energy at low frequencies. The cellular metamaterial shared the mechanical properties of honeycomb cores: having a density of $0.64g/cm^3$ but a yield strength of 21.2MPa in out-of-plane compression tests and an energy-absorption capability of $8.6J/cm^3$ in low-velocity impact tests.

Composites were prepared[182] by combining room-temperature vulcanized silicone rubber and a nanofiller such as multi-wall carbon nanotubes, nano-carbon black or graphite nanoplatelets; either alone or in a with molybdenum disulphide. The fillers ranged from 0-dimensional to 3-dimensional. The multi-wall carbon nanotubes, with a 1-dimensional tubular morphology and an aspect ratio of 65 exhibited good mechanical properties at a loading of 2 per hundred parts of rubber, followed by the nano-carbon black, the graphite nanoplatelets at 8 parts and MoS_2 at 2 parts. There was a synergistic effect upon the tensile strength and fracture strain due to hybrid fillers. The composites were used for piezo-electric energy generation.

Table 55. Structural parameters of XLiH₃

Parameter	MgLiH₃	CaLiH₃	SrLiH₃	BaLiH₃
a(Å)	3.76	4.29	4.64	5.04
B(GPa)	38.67	26.38	20.80	16.81
V(au³)	359.55	535.80	676.10	863.52

Natural-rubber composites have been made[183] from stearic acid-modified diatomaceous earth and carbon nanotubes, leading to markedly improved mechanical properties when compared with those of the plain rubber. The modified diatomaceous earth also offered better reinforcement than did the non-modified addition, due to the chemical surface-modification by stearic acid and consequently enhanced filler-rubber interaction. The addition of a small quantity of carbon nanotubes improved the mechanical properties and also improved the electrical properties and electromechanical sensitivity. The fracture toughness of the plain rubber was $9.74MJ/m^3$, and this was increased by some 484% to $56.86MJ/m^3$ by adding 40 parts per hundred grams of rubber of hybrid filler. Under similar testing conditions, the energy-harvesting efficiency of a composite which contained 57 parts per hundred of diatomaceous earth and 3 parts per hundred of carbon nanotubes was almost twice that of a composite which contained only 3 parts per hundred of carbon nanotubes. The greater energy-harvesting efficiency of modified diatomaceous earth-filled conductive composites was attributed to an increased dielectric behaviour.

The structural, opto-electronic, magnetic and mechanical properties of pristine M_3C_2 MXenes, where M = Zr or Mo, were determined. Energy versus volume plots confirmed chemical stability of their hexagonal crystal structures. Overlaid energy-states at the Fermi level in the band structures, and the density of states, confirmed that these MXenes were metallic[184]. Their mechanical stability was confirmed in terms of the Voigt-Reuss-Hill approximation. The Zr_3C_2 was antiferromagnetic, while Mo_3C_2 was paramagnetic and had a nett magnetic moment of $2\mu_B$. The Zr_3C_2 and Mo_3C_2 exhibited conductivities with peaks at 2.70 and 3.15/(fs), respectively.

Table 56. Calculated elastic constants of XLiH₃

Constant	MgLiH₃	CaLiH₃	SrLiH₃	BaLiH₃
C_{11}	68.79	47.07	41.63	36.91
C_{12}	22.47	16.18	9.86	4.97
C_{44}	11.57	7.90	5.95	3.88
B(GPa)	37.91	26.48	20.45	15.62
E(GPa)	40.47	27.48	23.17	22.13
G(GPa)	15.31	10.35	8.83	8.75
B/G	2.47	2.55	2.82	2.80
υ	0.32	0.32	0.31	0.31
T_m(K)	959.61	831.23	799.07	771.17

Dielectric elastomer generators comprise a dielectric elastomer film which is sandwiched between 2 flexible electrodes. They offer a high energy-density and a high energy-conversion efficiency and are useful for energy-harvesting. The impact of flexible electrode compliance upon the energy-harvesting stability and fatigue life of dielectric elastomer generators was investigated[185], indicating that the solution to good flexible electrode compliance is to reduce the difference in magnitude of the complex modulus and phase-angle between the flexible electrodes and dielectric elastomer. This could markedly reduce interfacial friction and extend the fatigue life of dielectric elastomer generators. With improved flexible-electrode compliance, the fatigue life and full-life energy density of the generator was increased by 20.3 times and 26.4 times, respectively, when compared with those of common carbon-based electrodes. A quantitative characterization method for the compliance of flexible electrodes was based upon the interfacial frictional resistance.

A study was made of the perovskite hydrides (table 55), XLiH₃, where X = Mg, Ca, Sr or Ba[186]. The compounds were stable in the cubic phase under atmospheric pressures and ambient temperatures. Their hydrogen-storage capacities were 8.76, 5.99, 3.07 and 2.03%, respectively. The SrLiH₃ had the greatest Debye temperature (344.41K). Analysis of the electronic band structure and density-of-states showed that the perovskite hydrides had a semiconducting nature, with indirect band-gaps of 2.66, 2.34, 1.94 and 1.37eV, respectively. The elastic constants, moduli and anisotropy were calculated. The Poisson

Energy-Harvesting Materials   Materials Research Forum LLC
Materials Research Foundations **177** (2025)   https://doi.org/10.21741/9781644903674

and Pugh ratios (table 56) indicated the compounds exhibited a ductile behaviour, with appreciable anisotropy.

Branched polyvinylidene fluoride nanofibres have been directly electrospun[187] with a diameter of less than 40nm by using tetrabutylammonium chloride at a relative humidity of 10%. The branched nanofibres exhibited good mechanical properties, a high crystallinity, a good β-phase content and a good electrical output.

The electrode materials make up most of the construction cost of microbial electrochemical systems. The low mechanical strength and poor electrochemical properties of common carbon-based materials limit the use of this technology. Polymer-based 3-dimensional honeycomb-structured materials with good mechanical properties have been used as supporting materials. Graphene, carbon nanotubes and polypyrrole have been chosen[188] as a surface-conductive coating layer for anodes. The introduction of the present choices increased the surface roughness, hydrophilicity, oxygen and nitrogen content and the electrochemically active surface area, while decreasing charge transfer, internal resistance and increasing extracellular electron transfer at electrode/microbe interfaces. The carbon nanotube anode system offered the best maximum power-density, 1700.7mW/m^2, among the 3 modified anode systems.

Moist-electric generators, based upon a polyhydroxyurethane vitrimer have been studied[189]. Unlike traditional polymer generators, the vitrimer moist-electric generators could provide high sustained voltage outputs under high-temperature and high-moisture conditions with no structural collapse or decomposition. Following doping with Ti_3C_2, vitrimer moist-electric generators could simultaneously convert energy from water and sun into electrical energy. The preparation of vitrimer moist-electric generators could be scaled up by using polyhydroxyurethane/cellulose paper composites. The resultant paper-based vitrimer moist-electric generators exhibited a high electrical output, together with good mechanical properties, a self-healing ability, plus high flexibility and transparency.

The bromination of poly(isobutylene-co-isoprene) rubber introduces 1 to 2mol% of bromide groups into the elastomer backbone and enables functionalisation. Three types of nucleophile reagent, pyridine, triphenylphosphine and imidazole, were introduced[190] into brominated poly(isobutylene-co-isoprene) rubber by nucleophile substitution. The nucleophile reagents bore 4 types of side-groups of methyl, ethyl, hydroxyl or vinyl form. The resultant ionic aggregates acted as physical cross-links and their size and density directly affected the mechanical reinforcement, the self-healing and the dynamic mechanical properties of the elastomers. Small polar imidazolyl/bromine pairs led to the highest degree of reinforcement, while 1-ethyl imidazole-modified rubber had the highest tensile strength (17.01MPa) and fracture elongation (1402%). Its self-healing efficiency was 63.7% following treatment at 140C for 0.5h. The inclusion of ionic clusters increased the relative permittivity of the elastomer and thus improved the energy-conversion efficiency.

Table 57. Characteristics of silicone rubber and nanoclay composites

Material	Cross-Links(mol/m^3)	E(MPa)	e_f(%)	V_b(V/μm)	Permittivity
SY0	-	0.87	280	19.1	2.73
SY2MT	-	0.48	170	18.3	3.54
SY4MT	-	0.24	220	21.3	3.14
B0	160	0.52	690	24.8	2.75
B2MT	117	0.35	1050	35.7	2.75
B4MT	84	0.17	1151	30.2	3.23

e_f: fracture elongation, V_b: breakdown voltage

The energy-harvesting ability and electro-mechanical properties of lead-free (1-x)(BaZr$_{0.2}$Ti$_{0.8}$)O$_3$-x(Ba$_{0.7}$Ca$_{0.3}$Ti)O$_3$ ceramics were determined[191]. A rhombohedral-orthorhombic-tetragonal phase-boundary region formed between 4/6 and 6/4 compositional ratios of (BaZr$_{0.2}$Ti$_{0.8}$)O$_3$ and (Ba$_{0.7}$Ca$_{0.3}$Ti)O$_3$. The Raman modes shifted toward lower frequencies with increasng Zr/Ca ratio and this attributed to asymmetrical Ti-O phonon vibrations which led to local disorder, to widening of the energy band and to a reduced Curie temperature. A large mechanical quality-factor of 556 was related to the hardening effect. A particularly high energy-conversion efficiency of 96% was found for a 3/7 (BaZr$_{0.2}$Ti$_{0.8}$)O$_3$/(Ba$_{0.7}$Ca$_{0.3}$Ti)O$_3$ ratio. The best electromechanical properties were found for a 5/5 (BaZr$_{0.2}$Ti$_{0.8}$)O$_3$/(Ba$_{0.7}$Ca$_{0.3}$Ti)O$_3$ ratio, with 500pC/N and 540pm/V. The good energy-harvesting characteristics included a voltage constant of 0.027Vm/N and a figure-of-merit under off-resonance conditions of 12.1 x 10^{-10}m^2/N for a 5/5 (BaZr$_{0.2}$Ti$_{0.8}$)O$_3$/(Ba$_{0.7}$Ca$_{0.3}$Ti)O$_3$ ratio. These excellent characteristics were attributed to the rhombohedral-orthorhombic-tetragonal phase boundary region, which comprised a low-energy barrier that facilitated polarization rotation and permitted increased electromechanical conversion.

Table 58. Key to material compositions

Designation	Base	Organo-Modified Montmorillonite(wt%)
SY0	Sylgard 186	0
B0	Silbione 4310	0
SY2MT	Sylgard 186	2
SY4MT	Sylgard 186	4
B4MT	Silbione 4310	4

Composites comprising $(Ba,Ca)(Zr,Ti)O_3$, reduced graphene oxide, CuO, Y_2O_3 and poly(vinylidene fluoride) exhibited great elasticity and flexibility and were incorporated into nanogenerators[192]. The effect of reduced graphene oxide upon the dispersion of the ceramics, the mechanical properties and the electrical behaviour was investigated, showing that the distribution of the ceramics in the presence of reduced graphene oxide was much more uniform. The elastic modulus and elongation-to-failure increased to 4.6MPa and 250%, respectively, after introducing the reduced graphene oxide. An average voltage of 1.36V and current of 35nA were achieved by merely tapping the nanogenerator.

The integration of polymer chains leads to changes in the mechanical properties of organolead halide perovskites and makes films softer; leading to a lower modulus and the ability to dissipate applied mechanical stimuli. This was exploited[193] in order to create a self-powered sensor having an operating range of up to 450kPa together with a linear response and high sensitivity over the whole range. The films offered an energy-harvesting density of $1.1W/m^2$.

A substrate-free energy-harvester which comprised ZnO nano-arrays was embedded in a polydimethylsiloxane elastomeric matrix[194]. This flexible device had an elastic modulus of 3.3MPa and could be stretched to 250% of its original length. It could generate an open-circuit voltage up to 9.2V.

Table 59. Calculated elastic constants of
cubic $CdLu_2S_4$ and $CdLu_2Se_4$

Constant	$CdLu_2S_4$	$CdLu_2Se4$
C_{11}	139.64	126.86
C_{12}	48.97	50.11
C_{44}	28.38	19.69
B(GPa)	79.19	75.69
G(GPa)	34.26	25.81
E(GPa)	89.84	69.52
B/G	2.31	2.93

A soft composite was prepared[195] by using 2 grades of platinum-catalyzed silicone rubber and organo-modified nanoclay (tables 57 and 58). They offered a low Young's modulus and a high dielectric permittivity, but without suffering from low dielectric or viscous losses, and while retaining a high dielectric breakdown strength. A The competition between partial inhibition of cross-linking, caused by a nanoclay surface modifier, and an increase in cross-links due to the nanoclay and associated filler stiffness, led to a decrease in the Young's modulus but without the decreasing mechanical stability and consequently the lifetime of the dielectric elastomers. On the other hand, a bimodal network of silica nanoparticles, long siloxane chains and aggregates of exfoliated nanoclays, decreased the viscoelasticy. An increase in dielectric permittivity occurred for any film which contained organo-modified montmorillonite, while dielectric losses remained very low. The electrical breakdown strength was also improved by organo-modified montmorillonite, over that of pristine materials. The synergistic effect of reducing elastic moduli, increasing dielectric permittivity and dielectric breakdown strength explained the improved performance.

Table 60. Mechanical properties of PDMS-SiO₂-TiO₂ composites

SiO$_2$(pph)	TiO$_2$(pph)	E(MPa)	Strength(MPa)	Strain(%)	Hardness*
0	0	0.63	0.31	90	62.0
28	0	1.97	1.24	128	83.0
-	10	0.23	0.32	160	46.5
28	2	1.42	3.22	312	72.6
28	6	1.13	2.79	313	74.4
28	10	1.37	3.91	406	77.7
28	20	0.97	3.16	363	81.4
28	30	1.32	4.89	589	84.0
28	50	1.02	4.53	635	86.0

*Shore A

The energy-harvesting ability of dielectric elastomers is closely related to their intrinsic electrical and mechanical properties. It was demonstrated[196] that poly(styrene-butadiene-styrene) compositions could be converted into a self-healing dielectric elastomer which offered a high permittivity and low dielectric loss, and could be deformed up to large strains. By using a one-step reaction at room temperature for 20min, methyl-3-mercaptopropionate could be grafted to poly(styrene-butadiene-styrene). The resultant material could be deformed up to a strain of 1000%, with a relative permittivity of 7.5. When incorporated into a dielectric actuator, it could offer 9.2% strain for an electric field of 39.5MV/m. It could also generate an energy-density of 11mJ/g by energy-harvesting. Following mechanical damage, the self-healed elastomer could recover 44% of its breakdown strength during energy harvesting.

First-principles studies, using density functional theory, were made[197] of the spinel compounds, CdLu$_2$X$_4$, where X was S or Se. The lattice parameters were consistent with experimental data, and Born's criteria confirmed mechanical stability. A Poisson coefficient greater than 0.26 and a Pugh ratio greater than 1.75 (table 59) reflected the ductile nature of the compounds. Use of a modified Becke-Johnson exchange potential predicted a direct band-gap of 2.08eV for CdLu$_2$S$_4$ and of 1.44eV for CdLu$_2$Se$_4$.

Table 61. Parameters of dynamo-mechanical curves for PDMS-SiO₂-TiO₂ composites

SiO$_2$(pph)	TiO$_2$(pph)	T$_g$(C)	T$_m$(C)	E'(Pa)$_{-140C}$	E'(Pa)$_{25C}$
0	0	-125	-43	9.60×10^8	2.53×10^6
28	0	-125	-41	1.36×10^9	5.11×10^6
-	10	-122	-42	2.88×10^9	4.65×10^5
28	2	-126	-40	2.17×10^9	1.63×10^6
28	6	-123	-40	4.15×10^9	1.73×10^6
28	10	-124	-40	3.96×10^9	1.20×10^6
28	20	-124	-38	4.99×10^9	1.61×10^6
28	30	-125	-38	6.47×10^9	2.05×10^6
28	50	-125	-37	5.33×10^9	2.58×10^6

E': storage modulus

Metallic bismuth offers a high volumetric capacity, a good operating potential and very reversible redox behaviour. On the other hand, it suffers from inferior long-term anodic stability and a melting-point of only 271C. An engineering innovation[198] permitted bismuth nano-units to be packaged into carbon sheaths by using NH₄Bi₃F₁₀. The high electrovalent Bi-F bond preserves bismuth thermal durability to above 400C. Bismuth-carbon hybrids comprised hollowed-out nanoscale features with accessible surface areas, smooth ionic transport channels and good mechanical properties. As-built anodes exhibited a good specific capacity and rate capability, with 53.9% capacity-retention at 20A/g. Nickel-bismuth cells offered a peak energy-density of 82.32Wh/kg and a peak power-density of /15.7kW/kg.

An easily scalable method was used[199] to prepare silicone composites that were based upon high molecular weight polydiorganosiloxane co-polymer and hydrophobized silica and titania nanoparticles (table 60). The host polymer was synthesized by bulk ring-opening co-polymerization of various substituted cyclosiloxanes. Composites which were created by the mechanical incorporation of fillers were cross-linked by radicalisation. Actuation measurements revealed the occurrence of displacements ranging from 0.04 to 5.09nm/Vmm, and impulse voltages ranging from 6 to 20V for a dynamic force of 0.1 to 1kgf. A slight decrease in the melting-point, and an increase in the storage modulus were revealed by dynamo-mechanical curves (table 61). The glass transition temperature was not affected by the fillers. The low dielectric and mechanical losses of the silicone/titania composites made them candidate materials for actuators. The dielectric constant of the composites increased with increasing filler content, while the flexibility was maintained. The energy-harvesting ability of the materials was promising.

A ferrofluid liquid spring has been used[200] to suspend a magnet array for the purpose of harvesting vibration energy. The ferrofluid liquid spring reduced the resonant frequency of a microfabricated electromagnetic energy harvester to about 340Hz, and 36nW could be delivered to a matched load under 7g acceleration.

Salinity-Harvesting

This is an example of large-scale recuperation of energy from natural phenomena. A tremendous amount of energy can theoretically be extracted due to the difference in salinity between sea- and fresh- water; an amount which is of the same order of magnitude as the energy generated by a high dam. A critical factor is the fresh-water supply which is offered by rivers flowing into the sea. A lot of energy is of course already generated by hydroelectric schemes, but salinity-gradient energy is a substantial additional resource. The two main methods which are chosen to extract salinity-gradient energy are pressure-retarded osmosis and reverse electrodialysis. Both involve the use of membranes, and a drawback is that membrane longevity is limited by their fouling. Pressure-retarded osmosis involves the flow of ion-free water through a membrane so as to produce electricity by using a turbine. Reverse electrodialysis uses membranes to control ion, rather than water, transport and an electrical current is generated directly by the flow of ions. Capacitive systems have also been considered which involve the alternating charging with brine and discharging of fresh water from materials which contain membranes. Pressure-retarded osmosis and reverse electrodialysis have largely been used to exploit natural salinity-gradient energy with the aid of sea water and river water. As a further bonus, the use of thermolytic solutions makes it possible to extract energy from waste heat.

Bio-inspired nacre-like layered MXene and poly(3,4-ethylenedioxythiophene) - poly(styrenesulfonate) composite membranes for the harvesting of salinity-gradient energy from an organic solvent were fabricated[201]. The membranes possessed good mechanical properties and exhibited great stability in the usual organic solvents. The power which was generated by the membrane attained 3216nW in a 2/0.001M methanol-LiCl solution, with a record power generation of 6926nW upon adding NaOH to the methanol-LiCl solution. The membrane exhibited excellent cation selectivity and fast ion-transport. These results were attributed to an increased interlayer spacing between the MXene layers and improved cation selectivity due to the insertion of poly(3,4-ethylenedioxythiophene) - poly(styrenesulfonate) chains and to an increased dissociation of negative charges by the NaOH. This ultra-stable 2-dimensional nanochannel membrane could harvest energy from waste organic solvents.

Pressure-retarded osmosis generally captures salinity-gradient energy by using semipermeable membranes which permit the transport of water from a low-concentration solution to a high-concentration solution. The maximum amount of energy which can be theoretically extracted during the reversible mixing of a dilute and saline solution is expected to range from 0.75 to 14.1kWh/m³ of the low-concentration stream. The development of pressure-retarded osmosis has long been impeded by the lack of a membrane which can permit an adequate flow. The bulky support layers of reverse-osmosis membranes lead to a marked internal concentration polarization which greatly reduces the water-flux and thus the power-density. Specially designed cellulose acetate membranes for forward osmosis can markedly reduce the effects of internal concentration polarization. Thin-film composite polyamide membranes might permit 5.7 to 10W/m² to be extracted.

The main barriers to the economic production of energy by using natural waters are the development of membranes which exhibit minimal internal concentration polarization and fouling. The streams have to be extensively pre-treated in order to prevent membrane-fouling. A pilot scheme which used pressure-retarded osmosis to extract energy from natural salinity gradients generated less than 1W/m² by using asymmetrical cellulose acetate membranes; lower than the target power-density of 5W/m². Pre-treatment of river water and sea water can itself require some 0.2 to 0.3kWh/m³ of energy. A scheme with 50% overall efficiency offered an extractable energy, resulting from the mixing of river water and sea water, of 0.3 to 0.4kWh/m³.

The advantage of using reverse electrodialysis to extract energy is that electricity is directly generated from salinity gradients. Brine and fresh water are brought together in arrays which comprise alternating anion-exchange and cation-exchange membranes which directly generate an electrochemical potential. In principle, water can be split by some 10 membrane-pairs, with each pair generating 0.1 to 0.2V. In practice, some 20 or more pairs are required. Water-splitting can release oxygen, chlorine and hydrogen and these have to be dealt with in various ways. Iron-based redox agents can enable the recycling of electrolyte solutions and thus avoid the need to split water. One such agent is ferrocyanide, but this can produce undesirable hydrogen cyanide gas. The iron can moreover precipitate at membranes and electrode surfaces, thus leading to fouling.

Reverse electrodialysis power-densities and energy efficiencies have been increased by improving membrane materials and design.

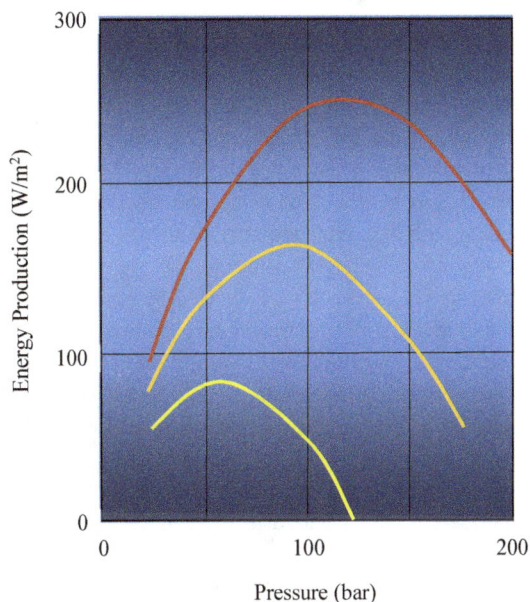

Figure 12. Theoretical membrane power density obtainable from an ammonia-CO₂ osmotic heat engine for 3.2 to 5.5M solutions under various applied hydraulic pressures.Red: 5.5M, orange: 4.6M, yellow: 3.2M

The power-density of a dialysis stack of 50 cell-pairs has attained $0.93\,W/m^2$ for a total membrane area at 3V of $0.5m^2$; that is, $93\,W/m^2$ with a planar cross-sectional area between the electrodes. A stack having a total area of $19m^2$ generated a nett power of $0.4\,W/m^2$, with an energy-efficiency of 18%. In the case of 25-membrane pair-stacks, optimization of the membrane materials and reduction of the spacing between membranes increased the power-density to between 1.2 and $2.2\,W/m^2$. The power-density was estimated to attain between 3.4 and $4\,W/m^2$ for reverse electrodialysis and $7.7\,W/m^2$ ($4kW/m^3$) for pressure-retarded osmosis when river-seawater interaction was involved.

For these orders of power density, 9W reactors with a comparable power-per-area could produce $0.16kW/m^3$ for reverse electrodialyis and $1.1kW/m^3$ for pressure-retarded osmosis.

A closed-loop system which can convert thermal energy into mechanical work or electricity is ideal for energy production[202]. In these systems, low- and high-concentration streams are used to generate energy via pressure-retarded osmosis or reverse electrodialysis, with thermal energy being used to separate the resultant mixed stream into low- and high-concentration streams that can be recycled. Closed-loop systems thus use salt solutions which can be separated using low-grade heat. Thermolytic solutions, involving ammonia and carbon dioxide are very promising (figure 12) and can generate power-densities of some $250W/m^2$ at 5M concentrations. Highly soluble ammonium salts can create a high osmotic pressure, are well-rejected by semipermeable membranes and can be separated from water and recycled using low-grade heat via simple distillation. A large quantity of energy can be extracted from the reversible mixing of a dilute stream with a ammonia and carbon dioxide solution. At 50C, the heat energy which is required to separate dilute ammonia and carbon dioxide solution into water (low-concentration stream) and ammonia plus carbon dioxide (which constitute a high-concentration stream) is $99kWh/m^3$ for a 1M solution and $165kWh/m^3$ for a 2M solution. Highly soluble salts such as sodium chloride can also be used in closed-loop heat engines with separation processes that use low-grade heat. Membrane distillation exploits the vapour-pressure gradient, created by the temperature difference across the membrane, to transport water vapour through a microporous hydrophobic membrane.

So osmotic energy, also known as blue energy, is a renewable source with zero emissions, but the necessary membranes for its harvesting require properties are difficult to create. Cartilage-like cation-selective composite membranes comprising aramid nanofibres and boron nitride nanosheets have made them possible[203] by using layer-by-layer assembly. Osmotic energy (figure 13) could then potentially be harvested by exploiting both salt-concentration gradient and pressure-driven streaming due to the high strength, toughness, chemical resistance and rapid ion-transport of the aramid/boron-nitride composites. Layered membranes constructed with molecular-scale precision exhibited both a high stiffness and tensile strength under repeated pressure-drops and salinity gradients. The total generated power-density over large areas was greater than $0.6W/m^2$ and could be maintained for up to 200h. The membranes could permit osmotic-energy harvesting between 0 and 95C, in pH-levels of between 2.8 and 10.8. The membranes were also resistant to oxidation. Another study[204] of these membranes revealed an average output power-density of $17.3W/m^2$ for more than 240h at a pH-level of 0. By exploiting the high-temperature properties of aramid, the output power-density was further increased to 77W/m² at 70C when applied to industrial wastewater. The results were an order-of-magnitude better than those for existing membranes. The improved efficiency of energy-harvesting was attributed to the high proton-selectivity of the present fibres.

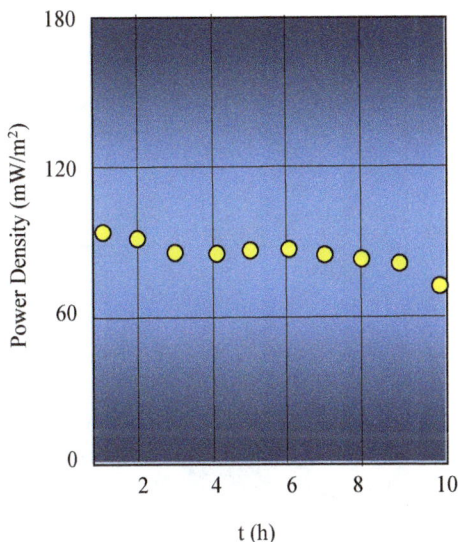

Figure 13. Osmotic-power density of aramid/boron-nitride membranes as a function of time for 0.01 to 0.5M NaCl gradient solutions

A coupling of solution-salinity and the mechanics of charged hydrogels offers another possibility for harvesting osmotic energy. Mechanical pressure is here applied in order to retard the swelling and de-swelling of hydrogels in saline solutions. The free energy of mixing is thereby converted into mechanical work. Mathematical modeling revealed[205] the influence of parameters such as the charge and elastic modulus of the hydrogels, the applied pressure and the solution salinity upon the energy conversion under various thermodynamic conditions. With optimum material design and control, the thermodynamic efficiency of the ideal process was predicted to be greater than 5% when applied to 10mM and 600mM NaCl solutions. The study of polystyrene sulfonate hydrogels confirmed the theoretically predicted trends, and a better than 10% thermodynamic efficiency for moderate-salinity sources, due to the swelling-strengthened mechanical properties of the gels.

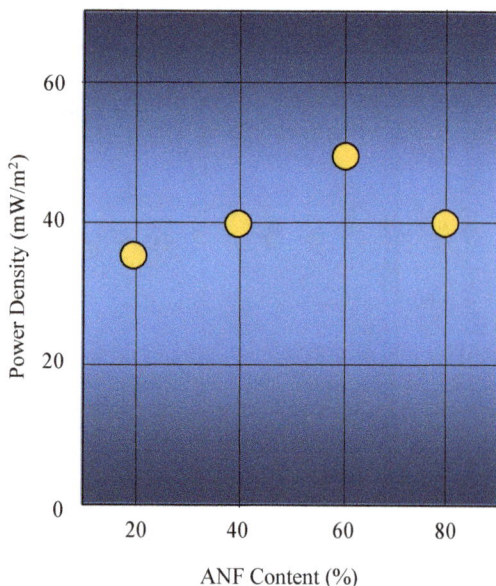

Figure 14. Output power-density of poly(sodium 4-styrene sulfonate)-C₃N₄/aramid nanofibre membranes for a 50-fold salinity-gradient, as a function of the aramid nanofibre content

Reverse electrodialysis tends to be limited by the channel-size and low charge-density. Thus graphitic carbon nitride, a graphite-layered polymer which possesses both a nanopore structure and a controllable surface charge, has been explored[206] as a highly-selective ion-exchange membrane. The carbon nitride membranes unfortunately still exhibit a low power-density. The graphitic C_3N_4 was functionalized with negatively-charged molecules such as polyacrylic acid and poly (sodium 4-styrenesulfonate) with nitrogen-containing groups. The functionalized materials were combined with anodized aluminium oxide membranes in order to form heterochannel membranes which could transport counter-ions and harvest osmotic energy. The surface charge-density of the functionalized material was improved, and this enhanced the output power-density. A heterochannel membrane which was functionalized with poly (sodium 4-styrenesulfonate) offered a power-density of $31mW/m^2$ for an effective area of $4.9mm^2$. Two-dimensional membranes were created by combining the present material with aramid nanofibres. These could maintain an output power-density of $50mW/m^2$ (figure 14).

Table 62. Power-density for carbon-
nanotube/polybutadiene membranes, versus pH

pH	Power-Density(W/m^2)
10	4.6
9	4.5
7	4.2
5	3.9
3	3.8

For the purpose of reverse electrodialysis, natural biomass membranes have been used[207]. The material can combine a yield strength of about 63.5MPa and a strain of some 2%. There are numerous nanoscale pores and negatively-charged -OH groups on the surface, and these provide minute nanofluidic channels for efficient cation trans-membrane transport. This imparts an excellent selectivity for cations during salinity-gradient energy conversion. The membranes offer osmotic-energy conversion with a power output-density of $22.2W/m^2$ in NaCl solutions; maintained at 98.53% for more than 6000s.

Carbon nanotubes offer an excellent conductivity and provide good channels for ion transport. As they cannot alone form membranes, they have been enclosed in hydroxyl-terminated polybutadiene matrices[208]. The nanotubes are first subjected to a plasma treatment which increases the surface charge density and transport capacity of the nanochannels. This improves the ion selectivity and energy conversion efficiency. Under seawater/fresh-water conditions (tables 62 to 64) such membranes offer a power-density of some $5.1W/m^2$, with a mechanical strength of 219MPa.

Table 63. Power-density for carbon-nanotube/polybutadiene membranes, versus time

Time (d)	Power-Density(W/m2)
1	4.2
3	4.1
5	3.8
7	3.7
9	3.7

Negatively-charged sodium carboxymethyl cellulose molecular chains have been introduced[209] into metallic graphene-like nanosheets in order to create nanofluidic materials. The cellulose chains greatly increase the number of negative charge sites in nanofluidic materials, thus increasing selectivity. They also increase the channel size, leading to a high cation flux. The composite materials markedly surpass pure graphene-like nanofluidics with regard to mechanical properties and produce a great improvement in osmotic energy conversion. Under seawater and river-water conditions, the composites can offer a power-density of up to $17.10 W/m^2$: an order-of-magnitude greater than that of pure graphene-type materials.

Table 64. Power-density for carbon-nanotube/polybutadiene membranes, versus concentration gradient

Concentration Gradient	Power-Density(W/m^2)
50	4.1
100	12.0
500	23.3

A tough zwitterionic gradient double-network hydrogel membrane which exhibits excellent biofouling resistance and cytocompatibility was proposed[210] for osmotic-energy harvesting. The membrane comprised negatively-charged 2-acrylamido-2-methylpropanesulfonic acid as a primary scaffolding network, and zwitterionic

Materials Research Forum LLC

https://doi.org/10.21741/9781644903674

sulfobetaine acrylamide as a secondary network. It was prepared via a 2-step photo-polymerization process which created a continuous-gradient double-network nano-architecture and greatly improved the mechanical properties. The gradient nano-architecture imparted to the hydrogel membrane an apparent ionic diode effect and space-charge governed transport properties. This aided directional ion transport. The membrane offered a power-density of $5.44W/m^2$ when mixing artificial seawater and river water. The output power could be boosted to $49.6W/m^2$ by mixing salt-lake water and river water.

A bio-inspired ultra-strong nanocomposite membrane was created, using layer-by-layer methods, which was based upon aramid nanofiber and graphene oxide. The material was intended for salinity-gradient energy-harvesting from organic solutions[211]. The membrane supported a mechanical stress of 688MPa and retained its integrity after soaking in organic solvents for 24h. When exposed to LiCl in methanol, a membrane-based device having a working area of $113mm^2$ produced a current and power of $28\mu A$ and $3140nW$, respectively. An assembly of 14 such cells could produce a voltage of up to 1.82V.

About the Author

Dr. Fisher has wide knowledge and experience of the fields of engineering, metallurgy and solid-state physics, beginning with work at Rolls-Royce Aero Engines on turbine-blade research, related to the Concord supersonic passenger-aircraft project, which led to a BSc degree (1971) from the University of Wales. This was followed by theoretical and experimental work on the directional solidification of eutectic alloys having the ultimate aim of developing composite turbine blades. This work led to a doctoral degree (1978) from the Swiss Federal Institute of Technology (Lausanne). He then acted for many years as an editor of various academic journals, in particular *Defect and Diffusion Forum*. In recent years he has specialized in writing monographs which introduce readers to the most rapidly developing ideas in the fields of engineering, metallurgy and solid-state physics. He is co-author of the widely-cited student textbook, *Fundamentals of Solidification*. Google Scholar credits him with 8687 citations and a lifetime h-index of 14.

References

[1] Nicholson, E.G., Adams, F.W., Patent GB17656, 1899.

[2] Nytko, B.J., MSc Dissertation, Naval Post-Graduate School (USA), 2010.

[3] Wiggins, E.B., Patent US1916873, 1933.

[4] Patel, V.A., Tailor, A.D., Patel, M.R., Journal of Mechanical and Civil Engineering, 11[4] 2014, 42. https://doi.org/10.9790/1684-11474244

[5] McShane, D.W., Patent GB2278645, 1994.

[6] The Sunday Times (UK), 18th December 1994.

[7] Holley, N., Patent US2014/0196446, 2014.

[8] Sen, A.K., Patent US6353270, 2002.

[9] Sen, A.K., Energy Conversion and Management, 38[7] 1997, 665. https://doi.org/10.1016/S0196-8904(96)00079-9

[10] Mohajer, N., Abdi, H., Nahavandi, S., Asia-Pacific Power and Energy Engineering Conference, 2015.

[11] Loverich, J.J., Kraige, D.R., Frank, J.E., Geiger, R.T., Patent WO2013/177589, 2013.

[12] Markus, D., Hayes, M., Patent WO2016/014118, 2016.

[13] Jia, Y., Doctoral Thesis, Cambridge University (UK), 2014.

[14] Htike, T.T., Soe, H.L., Moe, K.M., Iconic Research and Engineering Journals, 3[3] 2019, 529.

[15] Chicone, C., Feng, Z.C., European Journal of Physics, 29, 2008, 1115. https://doi.org/10.1088/0143-0807/29/5/024

[16] He J., Fan X., Mu J., Wang C., Qian J., Li X., Hou X., Geng W., Wang X., Chou X., Energy, 194, 2020, 116871. https://doi.org/10.1016/j.energy.2019.116871

[17] Seol M.L., Jeon S.B., Han J.W., Choi Y.K., Nano Energy, 31, 2017, 233-238. https://doi.org/10.1016/j.nanoen.2016.11.038

[18] Zhang H., Yang Q., Xu L., Li N., Tan H., Du J., Yu M., Xu J., Nano Energy, 125, 2024, 109521. https://doi.org/10.1016/j.nanoen.2024.109521

[19] Mi H.Y., Jing X., Zheng Q., Fang L., Huang H.X., Turng L.S., Gong S., Nano Energy, 48, 2018, 327-336. https://doi.org/10.1016/j.nanoen.2018.03.050

[20] Kulandaivel A., Potu S., Babu A., Madathil N., Velpula M., Rajaboina R.K., Khanapuram U.K., Nano Energy, 120, 2024, 109110. https://doi.org/10.1016/j.nanoen.2023.109110

[21] Vijayakanth T., Ram F., Praveenkumar B., Shanmuganathan K., Boomishankar R., Angewandte Chemie, 59[26] 2020, 10368-10373. https://doi.org/10.1002/anie.202001250

[22] Cui Y., Yang T., Luo H., Li Z., Jing X., International Journal of Mechanical Sciences, 279, 2024, 109523. https://doi.org/10.1016/j.ijmecsci.2024.109523

[23] Sahoo A., Paul T., Makani N.H., Maiti S., Banerjee R., Sustainable Energy and Fuels, 6[19] 2022, 4484-4497. https://doi.org/10.1039/D2SE00786J

[24] Roscow J., Zhang Y., Taylor J., Bowen C.R., European Physical Journal - Special Topics, 224[14-15] 2015, 2949-2966. https://doi.org/10.1140/epjst/e2015-02600-y

[25] Moss S.D., Payne O.R., Hart G.A., Ung C., Smart Materials and Structures, 24[2] 2015, 023001. https://doi.org/10.1088/0964-1726/24/2/023001

[26] Motora K.G., Wu C.M., Jose C.R.M., Rani G.M., Advanced Functional Materials, 34[22] 2024, 2315069. https://doi.org/10.1002/adfm.202315069

[27] He Y., Xiong J., Hu Y., Guo Z., Wang S., Mao J., Polymer, 320, 2025, 128075. https://doi.org/10.1016/j.polymer.2025.128075

[28] Li Z., Xu B., Han J., Tan D., Huang J., Gao Y., Fu H., Chemical Engineering Journal, 460, 2023, 141737. https://doi.org/10.1016/j.cej.2023.141737

[29] Li Z., Chen Y., Hang T., Xu C., Shen J., Li X., Zheng J., Wu Z., Journal of Materials Science and Technology, 204, 2025, 81-90. https://doi.org/10.1016/j.jmst.2024.02.076

[30] Chen K., Li Y., Yang G., Hu S., Shi Z., Yang G., Advanced Functional Materials, 33[45] 2023, 2304809. https://doi.org/10.1002/adfm.202304809

[31] Xu Y., Bai Z., Xu G., Nano Energy, 108, 2023, 108224. https://doi.org/10.1016/j.nanoen.2023.108224

[32] Zhao K., Lv H., Meng J., Song Z., Meng C., Liu M., Zhang D., ACS Omega, 7[22] 2022, 18816-18825. https://doi.org/10.1021/acsomega.2c01743

[33] Liu Y., Yiu C., Jia H., Wong T., Yao K., Huang Y., Zhou J., Huang X., Zhao L., Li D., Wu M., Gao Z., He J., Song E., Yu X., EcoMat, 3[4] 2021, e12123. https://doi.org/10.1002/eom2.12123

[34] Li L., Wang R., Fu Y., Jin Z., Chen J., Du H., Pan X., Wang Y.C., Advanced Functional Materials, 35[2] 2025, 2412324. https://doi.org/10.1002/adfm.202412324

[35] Zhou Z., Zhang Z., Du X., Zhang Q.L., Yang H., ACS Sustainable Chemistry and Engineering, 10[12] 2022, 3909-3919. https://doi.org/10.1021/acssuschemeng.1c08095

[36] Zhao K., Gao Z., Zhou J., Ye Y., Zhang J., Zhang C., Meng C., Zhang B., Chemical Engineering Journal, 500, 2024, 156709. https://doi.org/10.1016/j.cej.2024.156709

[37] Yang C., Wang Y., Wang Y., Zhao Z., Zhang L., Chen H., Nano Energy, 118, 2023, 109000. https://doi.org/10.1016/j.nanoen.2023.109000

[38] Chen T., Wei Q., Ma Y., Tang Y., Ma L., Deng S., Xu B., Nano Energy, 127, 2024, 109752. https://doi.org/10.1016/j.nanoen.2024.109752

[39] Hazarika A., Deka B.K., Jeong C., Park Y.B., Park H.W., Advanced Functional Materials, 29[31] 2019, 1903144. https://doi.org/10.1002/adfm.201903144

[40] Ou K., Wang M., Meng C., Guo K., Shariar Emon N., Li J., Qi K., Dai Y., Wang B., Composites Science and Technology, 255, 2024, 110732. https://doi.org/10.1016/j.compscitech.2024.110732

[41] Chen Z., Xu M., Zhou C., Hu Z., Du Z., Fu X., Song Y., jia Y., Wen X., Wang J., Cai G., Yang S., Yin X., Nano Energy, 132, 2024, 110361. https://doi.org/10.1016/j.nanoen.2024.110361

[42] Ye C., Dong S., Ren J., Ling S., Nano-Micro Letters, 12[1] 2020, 12. https://doi.org/10.1007/s40820-020-0380-z

[43] Kumar B., Latif M., Adil S., Kim J., Results in Engineering, 21, 2024, 101935. https://doi.org/10.1016/j.rineng.2024.101935

[44] Ruan H., Chen X., Lv C., Gu X., Zhou Z., Lu S., Li Y., Cellulose, 30[18] 2023, 11425-11437. https://doi.org/10.1007/s10570-023-05568-z

[45] Yang B., Chen X., Li Y., Ruan H., European Polymer Journal, 219, 2024, 113407. https://doi.org/10.1016/j.eurpolymj.2024.113407

[46] Li Z., Lu Y., Xiao D., Sun Y., Xu Y., Han J., Xu J., Xu B., Li C., Nano Energy, 135, 2025, 110630. https://doi.org/10.1016/j.nanoen.2024.110630

[47] Das S., Mallik M., Parida K., Bej N., Baral J., Applied Surface Science Advances, 23, 2024, 100626. https://doi.org/10.1016/j.apsadv.2024.100626

[48] Xu W., Wong M.C., Guo Q., Jia T., Hao J., Journal of Materials Chemistry A, 7[27] 2019, 16267-16276. https://doi.org/10.1039/C9TA03382C

[49] Wu Z., Wang P., Zhang B., Guo H., Chen C., Lin Z., Cao X., Wang Z.L., Advanced Materials Technologies, 6[1] 2021, 2000737. https://doi.org/10.1002/admt.202000737

[50] Zhai Q., Yang J., Qiao W., Qiao J., Gao H., Li Z., Wang P., Tang C., Xue Y., ACS Applied Polymer Materials, 4[5] 2022, 3236-3246. https://doi.org/10.1021/acsapm.1c01795

[51] Sintusiri J., Harnchana V., Amornkitbamrung V., Wongsa A., Chindaprasirt P., Nano Energy, 74, 2020, 104802. https://doi.org/10.1016/j.nanoen.2020.104802

[52] Ra Y., You I., Kim M., Jang S., Cho S., Kam D., Lee S.J., Choi D., Nano Energy, 89, 2021, 106389. https://doi.org/10.1016/j.nanoen.2021.106389

[53] Lin M.F., Parida K., Cheng X., Lee P.S., Advanced Materials Technologies, 2[1] 2017, 1600186. https://doi.org/10.1002/admt.201600186

Energy-Harvesting Materials
Materials Research Foundations **177** (2025)

Materials Research Forum LLC
https://doi.org/10.21741/9781644903674

[54] Deka B.K., Hazarika A., Lee S., Kim D.Y., Park Y.B., Park H.W., Nano Energy, 73, 2020, 104754. https://doi.org/10.1016/j.nanoen.2020.104754

[55] Chen Y., Zhang H., Xu C., Deng L., Yang Q., Zhang H., Xing J., Xie L., Nano Energy, 111, 2023, 108395. https://doi.org/10.1016/j.nanoen.2023.108395

[56] Hong S., Lee J., Do K., Lee M., Kim J.H., Lee S., Kim D.H., Advanced Functional Materials, 27[48] 2017, 1704353. https://doi.org/10.1002/adfm.201704353

[57] Sun C., Shi J., Wang X., Journal of Applied Physics, 108[3] 2010, 034309. https://doi.org/10.1063/1.3462468

[58] Maruyama K., Kawakami Y., Mori K., Kurita H., Shi Y., Jia Y., Narita F., Composites Science and Technology, 223, 2022, 109408. https://doi.org/10.1016/j.compscitech.2022.109408

[59] Zuo X., Chen L., Pan W., Ma X., Yang T., Zhang X., Micromachines, 11[12] 2020, 1072, 1-17. https://doi.org/10.3390/mi11121072

[60] Kumar A., Singh J., Advanced Theory and Simulations, 4[8] 2021, 2100156. https://doi.org/10.1002/adts.202100156

[61] Ennawaoui C., Hajjaji A., Samuel C., Sabani E., Rjafallah A., Najihi I., Laadissi E.M., Loualid E.M., Rguiti M., El Ballouti A., Azim A., Applied System Innovation, 4[3] 2021, 57. https://doi.org/10.3390/asi4030057

[62] García-Casas X., Ghaffarinehad A., Aparicio F.J., Castillo-Seoane J., López-Santos C., Espinós J.P., Cotrino J., Sánchez-Valencia J.R., Barranco Á., Borrás A., Nano Energy, 91, 2022, 106673. https://doi.org/10.1016/j.nanoen.2021.106673

[63] Han M., Wang H., Yang Y., Liang C., Bai W., Yan Z., Li H., Xue Y., Wang X., Akar B., Zhao H., Luan H., Lim J., Kandela I., Ameer G.A., Zhang Y., Huang Y., Rogers J.A., Nature Electronics, 2[1] 2019, 26-35. https://doi.org/10.1038/s41928-018-0189-7

[64] Koven R., Mills M., Gale R., Aksak B., IEEE Transactions on Ultrasonics, Ferroelectrics, and Frequency Control, 64[11] 2017, 1735-1743. https://doi.org/10.1109/TUFFC.2017.2739745

[65] Choi E.S., Kim H.C., Muthoka R.M., Panicker P.S., Agumba D.O., Kim J., Composites Science and Technology, 209, 2021, 108795. https://doi.org/10.1016/j.compscitech.2021.108795

[66] Lee S.H., Kim J., Sensors, 22[16] 2022, 6280. https://doi.org/10.3390/s22166280

[67] Stoykov S., Litak G., Manoach E., European Physical Journal: Special Topics, 224[14-15] 2015, 2755-2770. https://doi.org/10.1140/epjst/e2015-02587-3

[68] Shorakaei H., Shooshtari A., Journal of Intelligent Material Systems and Structures, 29[6] 2018, 1120-1138. https://doi.org/10.1177/1045389X17730919

[69] Leveque M., Douchain C., Rguiti M., Prashantha K., Courtois C., Lacrampe M.F., Krawczak P., International Journal of Polymer Analysis and Characterization, 22[1] 2017, 72-82. https://doi.org/10.1080/1023666X.2016.1233784

[70] Chen B., Jia Y., Narita F., Kurita H., Shi Y., Composites B, 273, 2024, 111274. https://doi.org/10.1016/j.compositesb.2024.111274

[71] Wen J., Mechanics of Advanced Materials and Structures, 31[26] 2024, 8085-8094. https://doi.org/10.1080/15376494.2023.2254290

[72] Pan D., Liang Y., Zhang Z., Wu Z., Mechanical Systems and Signal Processing, 224, 2025, 112013. https://doi.org/10.1016/j.ymssp.2024.112013

[73] Banerjee P., Dalela S., Balaji P.S., Murugan S., Kumaraswamidhas L.A., Acta Mechanica, 234[8] 2023, 3337-3359. https://doi.org/10.1007/s00707-023-03553-y

[74] Wang C.H., Jia X.D., Wang S., Yu G.X., China Journal of Highway and Transport, 35[7] 2022, 100-112.

[75] Li Z., Wang X., Jia M., Lu Y., Colloids and Surfaces A, 702, 2024, 135160. https://doi.org/10.1016/j.colsurfa.2024.135160

[76] Sun J., Guo H., Ribera J., Wu C., Tu K., Binelli M., Panzarasa G., Schwarze F.W.M.R., Wang Z.L., Burgert I., ACS Nano, 14[11] 2020, 14665-14674. https://doi.org/10.1021/acsnano.0c05493

[77] Zhang H., Jeong C.K., Shen Z., Wang J., Sun H., Jian Z., Chen W., Zhang Y., Composites B, 236, 2022, 109813. https://doi.org/10.1016/j.compositesb.2022.109813

[78] Zeng W., Tao X.M., Chen S., Shang S., Chan H.L.W., Choy S.H., Energy and Environmental Science, 6[9] 2013, 2631-2638. https://doi.org/10.1039/c3ee41063c

[79] Kulkarni N.D., Saha A., Kumari P., Polymer Composites, 45[7] 2024, 6264-6277. https://doi.org/10.1002/pc.28194

[80] Bhunia S., Karan S.K., Chowdhury R., Ghosh I., Saha S., Das K., Mondal A., Nanda A., Khatua B.B., Reddy C.M., Chem, 10[6] 2024, 1741-1754. https://doi.org/10.1016/j.chempr.2024.01.019

[81] Xue R., Huang Y., Zhang J., Wang T., Wu Y., Shi Q., Journal of Applied Polymer Science, 141[44] 2024, e56179. https://doi.org/10.1002/app.56179

[82] Kulkarni N.D., Kumari P., Journal of Polymer Research, 30[2] 2023, 79. https://doi.org/10.1007/s10965-023-03449-4

[83] He Q., Li X., Zhang H., Briscoe J., Advanced Functional Materials, 33[20] 2023, 2213918. https://doi.org/10.1002/adfm.202213918

[84] Stepancikova R., Olejnik R., Matyas J., Masar M., Hausnerova B., Slobodian P., Sensors, 24[4] 2024, 1275. https://doi.org/10.3390/s24041275

[85] Kumar V., Alam M.N., Yewale M.A., Park S.S., Polymers, 16[14] 2024, 2058. https://doi.org/10.3390/polym16142058

[86] Kurt P., Narayan B., Roscow J.I., Orhan S., Mechanics of Advanced Materials and Structures, 31[28] 2024, 10721-10734. https://doi.org/10.1080/15376494.2023.2295383

[87] Parvin N., Kumar V., Manikkavel A., Park S.S., Kumar Mandal T., Woo Joo S., Applied Surface Science, 613, 2023, 156078. https://doi.org/10.1016/j.apsusc.2022.156078

[88] Masiuchok O., Otulakowski Ł., Olszowska K., Chaber P., Foryś A., Buinova Y., Kolisnyk R., Myalska-Głowacka H., Szperlich P., Toroń B., Szeluga U., Godzierz M., Advanced Materials Technologies, 9[14] 2024, 2302040. https://doi.org/10.1002/admt.202302040

[89] Craciun F., Cordero F., Mercadelli E., Ilic N., Galassi C., Baldisserri C., Bobic J., Stagnaro P., Canu G., Buscaglia M.T., Dzunuzovic A., Petrovic M.V., Composites B, 263, 2023, 110835. https://doi.org/10.1016/j.compositesb.2023.110835

[90] Baraskar B.G., Kolekar Y.D., Thombare B.R., James A.R., Kambale R.C., Ramana C.V., Small, 19[37] 2023, 2300549. https://doi.org/10.1002/smll.202300549

[91] Khorshidi K., Rezaeisaray M., Karimi M., Acta Mechanica, 233[10] 2022, 4273-4293. https://doi.org/10.1007/s00707-022-03324-1

[92] Li X., Yuan C., Gao G., Zhou H., Zhang N., Yang X., Liu X., Sun H., Journal of Applied Polymer Science, 141[28] 2024, e55626. https://doi.org/10.1002/app.55626

[93] Maiti P., Sasmal A., Arunachalakasi A., Mitra R., ACS Applied Electronic Materials, 5[9] 2023, 4968-4983. https://doi.org/10.1021/acsaelm.3c00743

[94] Drdlik D., Marak V., Hadraba H., Chlup Z., Materials Letters, 355, 2024, 135424. https://doi.org/10.1016/j.matlet.2023.135424

[95] Parvin N., Kumar V., Park S.S., Mandal T.K., Joo S.W., Surfaces and Interfaces, 44, 2024, 103681. https://doi.org/10.1016/j.surfin.2023.103681

[96] Zhang H., Wu Y., Jin W., Chen W., Zhang Y., Materials Science in Semiconductor Processing, 155, 2023, 107260. https://doi.org/10.1016/j.mssp.2022.107260

[97] Perna G., Bonacci F., Caponi S., Clementi G., Di Michele A., Gammaitoni L., Mattarelli M., Neri I., Puglia D., Cottone F., Nanomaterials, 13[22] 2023, 2953. https://doi.org/10.3390/nano13222953

[98] Khazani Y., Rafiee E., Samadi A., Mahmoodi M., Colloids and Surfaces A, 687, 2024, 133537. https://doi.org/10.1016/j.colsurfa.2024.133537

[99] Wang R.C., Lin Y.C., Chen H.C., Lin W.Y., Nano Energy, 83, 2021, 105743. https://doi.org/10.1016/j.nanoen.2021.105743

[100] Xu M., Wu T., Song Y., Jiang M., Shi Z., Xiong C., Yang Q., Journal of Materials Chemistry C, 9[43] 2021, 15552-15565. https://doi.org/10.1039/D1TC03886A

[101] Wu T., Song Y., Shi Z., Liu D., Chen S., Xiong C., Yang Q., Nano Energy, 80, 2021, 105541. https://doi.org/10.1016/j.nanoen.2020.105541

[102] Sarafraz A.A., Roknizadeh S.A.S., Journal of Applied and Computational Mechanics, 3[2] 2017, 92-102.

[103] Rodrigues-Marinho T., Pereira N., Correia V., Miranda D., Lanceros-Méndez S., Costa P., ACS Applied Electronic Materials, 4[1] 2022, 287-296. https://doi.org/10.1021/acsaelm.1c01004

[104] Kumar V., Mandal T.K., Kumar A., Alam M.N., Parvin N., Joo S.W., Lee D.J., Park S.S., Express Polymer Letters, 16[9] 2022, 978-995. https://doi.org/10.3144/expresspolymlett.2022.71

[105] Ico G., Showalter A., Bosze W., Gott S.C., Kim B.S., Rao M.P., Myung N.V., Nam J., Journal of Materials Chemistry A, 4[6] 2016, 2293-2304. https://doi.org/10.1039/C5TA10423H

[106] Toroń B., Mistewicz K., Jesionek M., Kozioł M., Stróż D., Zubko M., Energy, 212, 2020, 118717. https://doi.org/10.1016/j.energy.2020.118717

[107] Cai J., Hu N., Wu L., Liu Y., Li Y., Ning H., Liu X., Lin L., Composites A, 121, 2019, 223-231. https://doi.org/10.1016/j.compositesa.2019.03.031

[108] Viet N.V., Zaki W., Umer R., Archive of Applied Mechanics, 90[12] 2020, 2715-2738. https://doi.org/10.1007/s00419-020-01745-9

[109] Liu Y.Z., Zhang H., Yu J.X., Huang Z.Y., Wang C., Sun Y., RSC Advances, 10[29] 2020, 17377-17386. https://doi.org/10.1039/D0RA01769H

[110] Zhao Q., Yang L., Chen K., Ma Y., Peng Q., Ji H., Qiu J., Composites Science and Technology, 199, 2020, 108330. https://doi.org/10.1016/j.compscitech.2020.108330

[111] Sultana A., Alam M.M., Ghosh S.K., Middya T.R., Mandal D., Energy, 166, 2019, 963-971. https://doi.org/10.1016/j.energy.2018.10.124

[112] Pan C.T., Yen C.K., Wu H.C., Lin L., Lu Y.S., Huang J.C.C., Kuo S.W., Journal of Materials Chemistry A, 3[13] 2015, 6835-6843. https://doi.org/10.1039/C5TA00147A

[113] Miao L., Song Y., Ren Z., Xu C., Wan J., Wang H., Guo H., Xiang Z., Han M., Zhang H., Advanced Materials, 33[40] 2021, 2102691. https://doi.org/10.1002/adma.202102691

[114] Yan J., Hou Y., Yu X., Zheng M., Zhu M., International Journal of Ceramic Engineering and Science, 3[4] 2021, 154-164. https://doi.org/10.1002/ces2.10088

[115] Jaita P., Jarupoom P., Sweatman D.R., Rujijanagul G., Journal of Asian Ceramic Societies, 9[3] 2021, 947-963. https://doi.org/10.1080/21870764.2021.1929739

[116] Yue Y., Hou Y., Zheng M., Fu J., Yan X., Zhu M., Materials Letters, 227, 2018, 21-24. https://doi.org/10.1016/j.matlet.2018.04.127

[117] Kurita H., Lohmuller P., Laheurte P., Nakajima K., Narita F., Additive Manufacturing, 54, 2022, 102741. https://doi.org/10.1016/j.addma.2022.102741

[118] Ren Z., Tang L., Zhao J., Zhang S., Liu C., Zhao H., Smart Materials and Structures, 31[10] 2022, 105001. https://doi.org/10.1088/1361-665X/ac798c

[119] Kim S.H., Thakre A., Patil D.R., Park S.H., Listyawan T.A., Park N., Hwang G.T., Jang J., Kim K.H., Ryu J., ACS Applied Materials and Interfaces, 13[17] 2021, 19983-19991. https://doi.org/10.1021/acsami.1c00922

[120] Barth S., Bartzsch H., Glöss D., Frach P., Modes T., Zywitzki O., Suchaneck G., Gerlach G., Microsystem Technologies, 22[7] 2016, 1613-1617. https://doi.org/10.1007/s00542-015-2787-x

[121] Krishna B., Chaturvedi A., Mishra N., Das K., Journal of Micromechanics and Microengineering, 28[11] 2018, 115013. https://doi.org/10.1088/1361-6439/aae10c

[122] Chen B., Jia Y., Narita F., Wang C., Shi Y., Composites B, 238, 2022, 109899. https://doi.org/10.1016/j.compositesb.2022.109899

[123] Malakooti M.H., Zhou Z., Sodano H.A., Nano Energy, 52, 2018, 171-182. https://doi.org/10.1016/j.nanoen.2018.07.051

[124] Roscow J.I., Pearce H., Khanbareh H., Kar-Narayan S., Bowen C.R., European Physical Journal: Special Topics, 228[7] 2019, 1537-1554. https://doi.org/10.1140/epjst/e2019-800143-7

[125] Nayak S., Chaki T.K., Khastgir D., Industrial and Engineering Chemistry Research, 53[39] 2024, 14982-14992. https://doi.org/10.1021/ie502565f

[126] Jacquelin E., Adhikari S., Friswell M.I., Smart Materials and Structures, 20[10] 2011, 105008. https://doi.org/10.1088/0964-1726/20/10/105008

[127] Kiriakidis G., Kortidis I., Cronin S.D., Morris N.J., Cairns D.R., Sierros K.A., Thin Solid Films, 555, 2014, 68-75. https://doi.org/10.1016/j.tsf.2013.05.149

[128] Topolov V.Y., Bowen C.R., Bisegna P., Sensors and Actuators, A, 229, 2015, 94-103. https://doi.org/10.1016/j.sna.2015.03.025

[129] Vatanabe S.L., Paulino G.H., Silva E.C.N., Computer Methods in Applied Mechanics and Engineering, 266, 2013, 205-218. https://doi.org/10.1016/j.cma.2013.07.003

[130] Wei J., Zhou Y., Wang Y., Miao Z., Guo Y., Zhang H., Li X., Wang Z., Shi Z., Energy, 265, 2023, 126398. https://doi.org/10.1016/j.energy.2022.126398

[131] Wang Z., Lv H., Gao Z., Song H., Chemical Engineering Journal, 498, 2024, 155789. https://doi.org/10.1016/j.cej.2024.155789

[132] Zhang C., Liu X., Han S., Wu M., Xiao R., Chen S., Chen G., Advanced Electronic Materials, 2025, in press.

[133] Qian W., Jia S., Yu P., Li K., Li M., Lan J., Lin Y.H., Yang X., Materials Today Physics, 49, 2024, 101589. https://doi.org/10.1016/j.mtphys.2024.101589

[134] Mustafa G.M., Saba S., Mahmood Q., Kattan N.A., Sfina N., Alshahrani T., Mera A., Mersal G.A.M., Amin M.A., Optical and Quantum Electronics, 55[6] 2023, 527. https://doi.org/10.1007/s11082-023-04666-3

[135] Li M., Chen H., Zhao J., Xia M., Qing X., Wang W., Liu Q., Lu Y., Luo M., Zhu X., Wang D., Advanced Composites and Hybrid Materials, 7[4] 2024, 106. https://doi.org/10.1007/s42114-024-00915-5

[136] Taheri A., MacFarlane D.R., Pozo-Gonzalo C., Pringle J.M., ChemSusChem, 11[16] 2018, 2788-2796. https://doi.org/10.1002/cssc.201800794

[137] Alqorashi A.K., Physica Scripta, 99[8] 2024, 085982. https://doi.org/10.1088/1402-4896/ad63e4

[138] Wan Y., Tan S., Li L., Zhou H., Zhao L., Li H., Han Z., Construction and Building Materials, 365, 2023, 130021. https://doi.org/10.1016/j.conbuildmat.2022.130021

[139] Tzounis L., Hegde M., Liebscher M., Dingemans T., Pötschke P., Paipetis A.S., Zafeiropoulos N.E., Stamm M., Composites Science and Technology, 156, 2018, 158-165. https://doi.org/10.1016/j.compscitech.2017.12.030

[140] Murtaza G., Javed A., Haseeb M., Aslam A.W., Rasul M.N., Hussain A., Materials Science in Semiconductor Processing, 190, 2025, 109366. https://doi.org/10.1016/j.mssp.2025.109366

[141] Khatun M.A., Mia M.H., Hossain M.A., Parvin F., Islam A.K.M.A., Journal of Physics and Chemistry of Solids, 196, 2025, 112381. https://doi.org/10.1016/j.jpcs.2024.112381

[142] Jin L., Greene G.W., MacFarlane D.R., Pringle J.M., ACS Energy Letters, 1[4] 2016, 654-658. https://doi.org/10.1021/acsenergylett.6b00305

[143] Lawongkerd J., Jongvivatsakul P., Vichai K., Yodprasert K., Pikkuwayo C., Prasittisopin L., Rungamornrat J., Keawsawasvong S., Sustainable and Resilient Infrastructure, 2025, in press.

[144] Park D., Kim M., Kim J., Applied Surface Science, 649, 2024, 159138. https://doi.org/10.1016/j.apsusc.2023.159138

[145] Mahmud S., Ali M.A., Hossain M.M., Uddin M.M., Vacuum, 221, 2024, 112926. https://doi.org/10.1016/j.vacuum.2023.112926

[146] Fang L., Chen C., Zhang H., Tu X., Wang Z., He W., Shen S., Wu M., Wang P., Zheng L., Wang Z.L., Materials Horizons, 11[6] 2024, 1414-1425. https://doi.org/10.1039/D3MH02228E

[147] Sun Y., Qin H., Zhang C., Wu H., Yin L., Liu Z., Guo S., Zhang Q., Cai W., Wu H., Guo F., Sui J., Nano Energy, 107, 2023, 108176. https://doi.org/10.1016/j.nanoen.2023.108176

[148] Jiang N., Qu M., Wang H., Bin Y., Zhang R., Tang P., Journal of Applied Polymer Science, 140[3] 2023, e53336. https://doi.org/10.1002/app.53336

[149] Yu C., Song Y.S., Macromolecular Research, 30[3] 2022, 198-204. https://doi.org/10.1007/s13233-022-0019-7

[150] Rehman S.U., Butt F.K., Ul Haq B., AlFaify S., Khan W.S., Li C., Solar Energy, 169, 2018, 648-657. https://doi.org/10.1016/j.solener.2018.05.006

[151] Yu C., Youn J.R., Song Y.S., Polymers for Advanced Technologies, 33[3] 2022, 700-709. https://doi.org/10.1002/pat.5419

[152] Lee C.H., Dharmaiah P., Kim D.H., Yoon D.K., Kim T.H., Song S.H., Hong S.J., ACS Applied Materials and Interfaces, 14[8] 2022, 10394-10406. https://doi.org/10.1021/acsami.1c23736

[153] Yu C., Kim H., Youn J.R., Song Y.S., ACS Applied Energy Materials, 4[10] 2021, 11666-11674. https://doi.org/10.1021/acsaem.1c02390

[154] Ur Rehman S., Butt F.K., Tariq Z., Ul Haq B., Lin G., Li C., Solar Energy, 185, 2019, 211-221. https://doi.org/10.1016/j.solener.2019.03.090

[155] Van Andel Y., Jambunathan M., Vullers R.J.M., Leonov V., Microelectronic Engineering, 87[5-8] 2010, 1294-1296. https://doi.org/10.1016/j.mee.2009.10.003

[156] Kwon S.C., Oh H.U., Sensors and Actuators A, 249, 2016, 172-185. https://doi.org/10.1016/j.sna.2016.08.030

[157] Malakooti M.H., Patterson B.A., Hwang H.S., Sodano H.A., Energy and Environmental Science, 9[2] 2016, 634-643. https://doi.org/10.1039/C5EE03181H

[158] Wen T., Ratner A., Jia Y., Shi Y., Composite Structures, 255, 2021, 112979. https://doi.org/10.1016/j.compstruct.2020.112979

[159] Jaaoh D., Putson C., Muensit N., Composites Science and Technology, 122, 2016, 97-103. https://doi.org/10.1016/j.compscitech.2015.11.020

[160] Shishesaz M., Shirbani M.M., Sedighi H.M., Hajnayeb A., Journal of Sound and Vibration, 425, 2018, 149-169. https://doi.org/10.1016/j.jsv.2018.03.030

[161] Zhang L., Lu J., Takei R., Makimoto N., Itoh T., Kobayashi T., Review of Scientific Instruments, 87[8] 2016, 085005. https://doi.org/10.1063/1.4960959

[162] Yurchenko D., Val D.V., Lai Z.H., Gu G., Thomson G., Smart Materials and Structures, 26[10] 2017, 105001. https://doi.org/10.1088/1361-665X/aa8285

[163] Ando B., Baglio S., Bulsara A.R., Marletta V., Ferrari V., Ferrari M., IEEE Sensors Journal, 15[6] 2015, 3209-3220. https://doi.org/10.1109/JSEN.2014.2386392

[164] Kluger J.M., Sapsis T.P., Slocum A.H., Journal of Sound and Vibration, 341, 2015, 174-194. https://doi.org/10.1016/j.jsv.2014.11.035

[165] Jin-Yang L.I., Zhu S., Structural Control and Health Monitoring, 29[12] 2022, e3120.

[166] Li Z., Jin H., Sun B., Railway Standard Design, 68[9] 2024, 53-59.

[167] Kim J., Choi Y., Jang H., Jiong S., Chen X., Seo B., Choi W., Advanced Materials, 36[47] 2024, 2411248. https://doi.org/10.1002/adma.202411248

[168] Liu W., Lan Y., Li H., Liu C., Dufresne A., Fu L., Lin B., Xu C., Huang B., International Journal of Biological Macromolecules, 286, 2025, 138229. https://doi.org/10.1016/j.ijbiomac.2024.138229

[169] Kim S., Choi S.J., Zhao K., Yang H., Gobbi G., Zhang S., Li J., Nature Communications, 7, 2016, 10146. https://doi.org/10.1038/ncomms10146

[170] Bao D., Wen Z., Shi J., Xie L., Jiang H., Jiang J., Yang Y., Liao W., Sun X., Journal of Materials Chemistry A, 8[27] 2020, 13787-13794. https://doi.org/10.1039/D0TA03215H

[171] Niu Z., Qi S., Shuaib S.S.A., Züttel A., Yuan W., Composites Science and Technology, 226, 2022, 109538. https://doi.org/10.1016/j.compscitech.2022.109538

[172] Xiao T., Zhao J., Geng L., Wang Z., Qiao W., Liu C., Chemical Engineering Journal, 477, 2023, 147068. https://doi.org/10.1016/j.cej.2023.147068

[173] Ayyaz A., Abaid Ullah M., Zaman M., Alkhaldi N.D., Mahmood Q., Boukhris I., Al-Buriahi M.S., Al-Anazy M.M., International Journal of Hydrogen Energy, 102, 2025, 1329-1339. https://doi.org/10.1016/j.ijhydene.2025.01.117

[174] Naeem H., Ullah M.A., Hussain A., Sandali Y., Usman Z., Rizwan M., International Journal of Hydrogen Energy, 105, 2025, 203-213. https://doi.org/10.1016/j.ijhydene.2025.01.318

[175] Wu W., Dong Z., Wang Y., Zang W., Jiang Y., Ning N., Tian M., ACS Sustainable Chemistry and Engineering, 12[14] 2024, 5447-5458. https://doi.org/10.1021/acssuschemeng.3c07155

[176] Rizwan M., Ali S.S., Sabahat U., Sana M., Zahid U., Ullah M.A., Chinese Journal of Physics, 86, 2023, 418-430. https://doi.org/10.1016/j.cjph.2023.11.004

[177] Manzoor M., Dixit A., Waqas Iqbal M., Sadique I., Alotaibi N.H., Mohammad S., Sharma R., Ur Rehman I., Ismayilova N.A., Materials Science and Engineering: B, 310, 2024, 117710. https://doi.org/10.1016/j.mseb.2024.117710

[178] Yu Y., Shi Y., Kurita H., Jia Y., Wang Z., Narita F., Composites A, 172, 2023, 107587. https://doi.org/10.1016/j.compositesa.2023.107587

[179] Jiang H., Zhang R., Liu K., Luo Y., Peng Z., Ye S., Qin Y., Wu X., Gao C., Liu Y., Xu D., Xu W., Nano Energy, 132, 2024, 110407. https://doi.org/10.1016/j.nanoen.2024.110407

[180] Blancas-Flores J.M., Morales-Rivera J., Rocha-Ortiz G., Ahuactzi I.F.H., Cabrera-Chavarria J.J., Andrade-Melecio H.A., Astudillo-Sanchez P.D., Antolín-Cerón V.H., International Journal of Renewable Energy Development, 13[6] 2024, 1162-1174. https://doi.org/10.61435/ijred.2024.60664

[181] Xiao Z., Gao P., He X., Qu Y., Wu L., Materials and Design, 241, 2024, 112912. https://doi.org/10.1016/j.matdes.2024.112912

[182] Kumar V., Manikkavel A., Kumar A., Alam M.N., Hwang G.J., Park S.S., Journal of Vinyl and Additive Technology, 28[4] 2022, 813-827. https://doi.org/10.1002/vnl.21930

[183] Alam M.N., Kumar V., Jung H.S., Park S.S., Polymers, 15[17] 2023, 3612. https://doi.org/10.3390/polym15173612

[184] Khalil R.M.A., Shah M.H., Hussain M.I., Alotaibi N.H., Mohammad S., Hussain F., Meeladi G., Physica B, 685, 2024, 416016. https://doi.org/10.1016/j.physb.2024.416016

[185] Yao J., Zang W., Wang Y., Yu B., Jiang Y., Ning N., Tian M., ACS Applied Materials and Interfaces, 16[9] 2024, 11595-11604. https://doi.org/10.1021/acsami.3c19158

[186] Ayyaz A., Alkhaldi H.D., Saidi S., Albalawi H., Zayed O., Al-Daraghmeh T.M., Mahmood Q., Alqorashi A.K., Materials Science in Semiconductor Processing, 186, 2025, 109020. https://doi.org/10.1016/j.mssp.2024.109020

[187] Zaarour B., Zhu L., Jin X., Polymers for Advanced Technologies, 31[11] 2020, 2659-2666. https://doi.org/10.1002/pat.4992

[188] Li J., Chen D., Liu G., Li D., Tian Y., Feng Y., Water Research, 218, 2022, 118429. https://doi.org/10.1016/j.watres.2022.118429

[189] Feng Z., Zhao W., Yang Z., Deng Y., Yang T., Ni Y., Journal of Materials Chemistry A, 10[21] 2022, 11524-11534. https://doi.org/10.1039/D1TA11029B

[190] Hu X., Zhang R., Wemyss A.M., Elbanna M.A., Heeley E.L., Arafa M., Bowen C., Wang S., Geng X., Wan C., Materials Advances, 3[10] 2022, 4213-4226. https://doi.org/10.1039/D2MA00124A

[191] Bijalwan V., Erhart J., Spotz Z., Sobola D., Prajzler V., Tofel P., Maca K., Journal of the American Ceramic Society, 104[2] 2021, 1088-1101. https://doi.org/10.1111/jace.17497

[192] Wu Y., Ma F., Qu J., Qi T., Materials Letters, 231, 2018, 20-23. https://doi.org/10.1016/j.matlet.2018.07.102

[193] Saraf R., Tsui T., Maheshwari V., Journal of Materials Chemistry A, 7[23] 2019, 14192-14198. https://doi.org/10.1039/C9TA03982A

[194] Jin C., Dong L., Xu Z., Closson A., Cabe A., Gruslova A., Jenney S., Escobedo D., Elliott J., Zhang M., Hao N., Chen Z., Feldman M.D., Zhang J.X.J., Advanced Materials Interfaces, 8[10] 2021, 2100094. https://doi.org/10.1002/admi.202100094

[195] Boccalero G., Jean-Mistral C., Castellano M., Boragno C., Composites B, 146, 2018, 13-19. https://doi.org/10.1016/j.compositesb.2018.03.021

[196] Ellingford C., Zhang R., Wemyss A.M., Zhang Y., Brown O.B., Zhou H., Keogh P., Bowen C., Wan C., ACS Applied Materials and Interfaces, 12[6] 2020, 7595-7604. https://doi.org/10.1021/acsami.9b21957

[197] Zareef F., Rashid M., Ahmadini A.A.H., Alshahrani T., Kattan N.A., Laref A., Materials Science in Semiconductor Processing, 127, 2021, 105695. https://doi.org/10.1016/j.mssp.2021.105695

[198] Yao J., Li L., Li N., Jiang J., Wang Y., Zhu J., Materials Chemistry Frontiers, 4[4], 2020, 1249-1255. https://doi.org/10.1039/D0QM00017E

[199] Cazacu M., Ignat M., Racles C., Cristea M., Musteata V., Ovezea D., Lipcinski D., Journal of Composite Materials, 48[13] 2014, 1533-1545. https://doi.org/10.1177/0021998313488148

[200] Wang Y., Zhang Q., Zhao L., Kim E.S., Proceedings of the IEEE International Conference on Micro Electro Mechanical Systems, 2015, 122-125. https://doi.org/10.1109/MEMSYS.2015.7050901

[201] Chen Y., Fang M., Ding S., Liu Y., Wang X., Guo Y., Sun X., Zhu Y., ACS Applied Materials and Interfaces, 14[20] 2022, 23527-23535. https://doi.org/10.1021/acsami.2c04307

[202] Logan B.E., Elimelech M., Nature, 488, 2012, 313-319. https://doi.org/10.1038/nature11477

[203] Chen C., Liu D., He L., Qin S., Wang J., Razal J.M., Kotov N.A., Lei W., Joule, 4[1] 2020, 247-261. https://doi.org/10.1016/j.joule.2019.11.010

[204] Chen C., Yang G., Liu D., Wang X., Kotov N.A., Lei W., Advanced Functional Materials, 32[1] 2020, 2102080. https://doi.org/10.1002/adfm.202102080

[205] Zhang S., Lin S., Zhao X., Karnik R., Journal of Applied Physics, 128[4] 2020, 044701. https://doi.org/10.1063/5.0013357

[206] Chen Y., Qin Y., Yang J., Zhang H., Yang X., Wei L., Journal of Electroanalytical Chemistry, 952, 2024, 117967. https://doi.org/10.1016/j.jelechem.2023.117967

[207] Zheng X., Jia M., Yuan Z., Ma X., Teng C., Kong B., ACS Applied Materials and Interfaces, 16[51] 2024, 70618-70625. https://doi.org/10.1021/acsami.4c17932

[208] Lin C., Hao J., Zhao J., Hou Y., Ma S., Sui X., Journal of Colloid and Interface Science, 654, 2024, 840-847. https://doi.org/10.1016/j.jcis.2023.10.084

[209] Huang S., Liu T., Xin W., He X., Wan S., Yang C., Zhao J., Shi L., Zhou T., Wen L., Journal of Membrane Science, 718, 2025, 123678. https://doi.org/10.1016/j.memsci.2024.123678

[210] Huang K.T., Hung W.H., Su Y.C., Tang F.C., Linh L.D., Huang C.J., Yeh L.H., Advanced Functional Materials, 33[19] 2023, 2211316. https://doi.org/10.1002/adfm.202211316

[211] Chen C., Liu D., Yang G., Wang J., Wang L., Lei W., Advanced Energy Materials, 10[18] 2020, 1904098. https://doi.org/10.1002/aenm.201904098